£30

MI

A History of the Burmah Oil Company

T. A. B. CORLEY

A History of the Burmah Oil Company
1886–1924

HEINEMANN: LONDON

William Heinemann Ltd
10 Upper Grosvenor Street, London W1X 9PA

LONDON MELBOURNE TORONTO
JOHANNESBURG AUCKLAND

First published 1983
© The Burmah Oil Company 1983
ISBN 434 14520 3

Photoset in Great Britain by
Rowland Phototypesetting Ltd., Bury St Edmunds, Suffolk
and printed by St Edmundsbury Press,
Bury St Edmunds, Suffolk

'I confess with some regret that none of the men at present on our books is a Scotchman. Prejudice apart, there is something in our country's air and hard-bitten natural conditions which seems to produce men who "take hold" in our Empire and in every phase of its work. Our own business has been built up by them also, and I should like if we could to maintain the useful natural leaven in our staff.'

R. I. Watson (later Managing Director) to (Sir) John Cargill (Chairman), 23 December 1918

Contents

List of Plates

Author's Note

Some of the themes in the present book have been analysed by the author in the following articles: "Communications, Entrepreneurship and the Managing Agency System; The Burmah Oil Company 1886–1928", a paper given to the Business History Seminar, Polytechnic of Central London, May 1979 and summarized in *Business History*, XXII (1980), pp. 104–5, and "Strategic Factors in the Growth of a Multinational Enterprise; The Burmah Oil Company 1886–1928", in M. C. Casson (ed.) *The Growth of International Business* (1983), pp. 214–35.

I am sorry that two related studies were published too late to receive attention here: R. W. Ferrier, *The History of the British Petroleum Company, Volume I, The Developing Years 1901–1932* (1982) and G. G. Jones, *The State and the Emergence of the British Oil Industry* (1981).

Authors of books of this kind have to rely greatly on the willingness of people to provide reminiscences, hunt out old papers and go to much trouble in seeking help from others. I have received heart-warming responses from all but a very few of those approached, and I am more than grateful for the replies by correspondence, as for the courteous reception and kind hospitality extended to me. Even when the information gleaned does not appear directly, it has often provided valuable background material.

And as to individual acknowledgments. My thanks are due especially to John E. Harvey, former Public Affairs Director, for his steady encouragement of the history project during some difficult years for the company after 1974, and for his ready and often inspired suggestions for improving the manuscript during the revision process, and to H. H. (Oliver) Twist, former Public Relations Officer, who first assembled and then annotated many of the surviving archives and built up an unrivalled knowledge of the company's early days. His help and advice have been sorely missed since his retirement abroad.

By convention, existing company employees are not thanked by name, since it is often invidious to choose between so many. Even so, my particular thanks go to the Director of Public Affairs Richard Dixon, to David Prockter, Brian Jones, Peter Perry and Norman Wells at the Swindon head office, to Eileen Murphy and Gerry Booth in the London office, to Eric Drummond at the Glasgow office and to A. M. Boks and staff of Castrol Nederland. To all others I express my sincere gratitude for assistance and interest over the past ten years. Of those no longer with the company, a number of retired board members have helped, namely the late Sir William (WEVA) Abraham, the late W. G. Eadie, W. P. G. Maclachlan, J. F. Strain and H. R. Tainsh. Among former employees, I am also grateful to Tony Gowan, George Paterson, Dr A. M. Robertson and Carol Thomson for unfailing help. G. A. F. Grindle, formerly of the British Burmah Petroleum Company, undertook some very important preliminary research in the Public Record Office and India Office Library and freely contributed his extensive knowledge of oil history in Burma by commenting on some of the letter books.

Relatives of some of the main characters have generously devoted time and effort to answering my queries, including: for the Cargill family, the late Mrs Alison Greenlees (daughter of Sir John Cargill), the late W. D. Cargill Thompson (who shared his detailed researches into Glasgow's maritime history and Finlay Fleming & Co.'s past) and D. C. Todd; for the Finlay family, the late Lady (C. K.) Finlay and her children C. K. Finlay, Vivian Stuart (Mrs C. W. Mann) and Mrs John Ward, and also Mrs Penelope Alston and Miss Wendy Day; for H. S. Ashton, the late Sir Hubert Ashton and Lady Ashton; for some lively reminiscences by W. G. Corfield, his daughter Mrs Hereward Swallow; for M. T. Fleming, Captain R. Mackenzie; for the (younger) Galbraith family, Mrs D. Galbraith and Dr T. Russell; for the Greenway family the Hon. Mrs Dalzell

Hunter (the first Lord Greenway's daughter) and Mrs John Trechman; Lady (Adam) Ritchie; for C. W. Wallace, his grandsons C. B. Carr and Brigadier R. M. Carr; and for R. I. Watson, his son Neil G. Watson. I am also grateful for conversations about the Burma of the past with F. S. V. Donnison and Mrs V. Rhodes James.

Crown copyright material is reproduced by permission from the following sources: the Public Record Office, the India Office Library, the Library of H. M. Customs and Excise, and the Naval Historical Library, Ministry of Defence. At the Customs and Excise, the Librarian E. A. Carson gave every assistance, as did Captain M. Beeching R.N. of the Naval Historical Branch and the staff of the Mitchell Library, Glasgow. The Librarian of the London Library, S. Gillam, kindly allowed me to consult Sir Arnold Wilson's correspondence and to quote from it. At Price's Patent Candle Company F. Dixon was very helpful, as was C. E. Mold by correspondence. For the Irrawaddy Flotilla Company A. G. McCrae – author, with Alan Prentice, of *Irrawaddy Flotilla* (1978) – readily answered a number of queries. Dr Joseph Needham of Cambridge lent his great knowledge to the problem of the thirteenth-century Chinese traveller (see Introduction). At the University of Reading, Mrs Margaret Lewis speedily and expertly typed some of the earlier drafts.

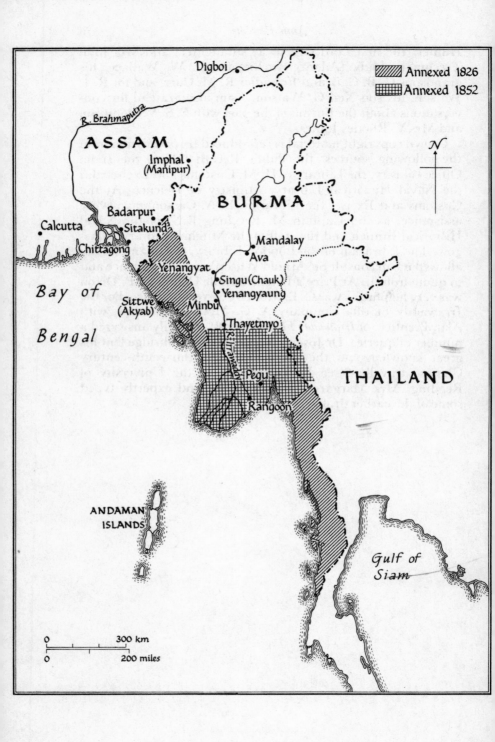

Annexed 1826
Annexed 1852

Digboi

R. Brahmaputra

ASSAM

Imphal
(Manipur)

BURMA

Badarpur
Calcutta
Sitakund
Chittagong

Mandalay
Ava

Yenangyat
Singu (Chauk)
Yenangyaung

Bay of
Bengal

Sittwe
(Akyab)
Minbu

Thayetmyo

R. Irrawaddy

Pegu

THAILAND

Rangoon

ANDAMAN
ISLANDS

Gulf of
Siam

0   300 km
0   200 miles

The Burmah Oil Company and its Forebears

	UK Agency	Rangoon Agency (Burma)	Calcutta Agency (India)
Syndicate of East India Merchants, Rangoon, c.1858	—	—	—
The Burmese Oil Distillery Ltd, 1864–8 Chairman: John Ogle	Head Office, London	(not known)	—
The Rangoon Oil Company Ltd, 1871–6 Chairman: James Galbraith	Galbraith Reid & Co., Glasgow	Galbraith Dalziel & Co.	Wiseman Mitchell Reid & Co.
David Sime Cargill (sole proprietor) 1876–86	Milne & Co., Glasgow	Finlay Fleming & Co.	Mitchell Reid & Co.
The Burmah Oil Company Ltd, 1886 (reconstructed, 1902) Chairman: D. S. Cargill	Milne & Co., Glasgow and London	Finlay Fleming & Co.	Mitchell Reid & Co. (to 1891) Shaw Wallace & Co. (1891–1902)
The Burmah Oil Company Ltd, 1902 Chairmen: D. S. Cargill (d. 1904) (Sir) John T. Cargill (1904–43) Managing Directors: Kirkman Finlay (1900–3) (Manager: James Hamilton, 1905–19) R. I. Watson (1920–43)	Milne & Co., Glasgow and London (1902–6) Company's own Glasgow and London Offices thereafter	Finlay Fleming & Co. (to 1928)	Shaw Wallace & Co. (to 1928)

Introduction

Of all Britain's oil enterprises, the Burmah Oil Company is the oldest, with its antecedents stretching well back before the acknowledged birth date of the industry – Edwin Drake's discovery of Pennsylvanian well-oil in 1859. Among the manifold changes the company has since encountered, none is more striking than the transformation in the role of oil itself from a source of light to a source of power. When Burmah Oil was registered as a company in 1886, all its saleable products were earmarked for lighting, heating and lubricating purposes. By 1924, when the present volume ends, it not merely supplied two-thirds of mainland India's kerosene requirements, but also provided important sources of energy such as petrol (i.e. motor spirit), as well as quantities of fuel oil for the Royal Navy under the first long-term fuel oil contract the Admiralty ever negotiated.

Despite performing these vital tasks, the company kept well away from the glare of publicity at home. As its principal markets were in the Indian empire – which included the province of Burma – the average Briton knew nothing of its products, apart from the odd observant public schoolboy who noted when cleaning his Corps rifle that the pull-throughs were impregnated with Rangoon oil. Even its petrol sold in Britain bore the brand name of a rival company which had purchased that petrol in bulk. Its dividends, which remained consistently good year after year, sustained numerous (mainly Scottish) widows and orphans; yet annual meetings, or news items about the company, hardly ever made the columns of the popular press.

I

Similarly, the red sandstone head office in Glasgow – for it was, and remains, a Scottish-registered company – was distinguishable from hundreds of that city's other corporate head offices only by its superior architectural merit. By the early 1920s, the London office was discreetly tucked away on one floor of Sir Edwin Lutyens's imposing Britannic House in the City, headquarters of a much larger and truly more glamorous enterprise, the Anglo-Persian Oil Company, now British Petroleum. The observant schoolboy might have inferred that Burmah Oil was an obscure offshoot or poor relation of Anglo-Persian. He would have been surprised to learn that the smaller company had not only been the parent of the larger, but that, together with the British government, it long held the bulk of Anglo-Persian's ordinary shares: Burmah Oil's holding was disposed of as recently as 1975. The astonishing story of its involvement in the Persian operations and the eventful relationship with Anglo-Persian is a major theme in the narrative from 1904 onwards.

TABLE I
Leading world oil firms, 1924
(by issued capital: £mn equivalent)

Standard Oil of New Jersey (Esso/Exxon)	158.1
Royal Dutch – Shell	63.8
Standard Oil of California (Socal/Chevron)	52.1
Standard Oil of New York (Mobil)	51.0
Standard Oil of Indiana (Amoco)	50.1
Texas Corporation (Texaco)	36.8
Gulf Oil Corporation	24.4
Anglo-Persian Oil Company (British Petroleum)	19.5
Vacuum Oil Company (Mobil)	13.9
Burmah Oil Company	9.2

Source: W. E. Skinner, *The Oil and Petroleum Year Book* (1925)

The Burmah Oil Company began modestly enough in 1886 and, as Table 1 shows, even in 1924 it was considerably smaller than the leading firms in the world oil industry. Its total production in 1886 constituted only one-tenth of 1 per cent of the world's oil needs, and did not reach even 1 per cent by 1924. Yet its influence soon outstripped its relatively lowly rank, for it was on the doorstep of what was then the largest market for oil products in the old world outside Europe, the subcontinent of India. It therefore came under the baleful scrutiny of the "giants" such as Standard Oil, Royal Dutch and Shell. Again and again up to 1924 and beyond, the interplay between the three or four rival companies was of vital concern to the Burmah Oil directors.

More significantly for the future, in the whole of the globe-circling British empire, Burmah Oil was the only wholly British oil producer of any consequence. To this company, therefore, the Admiralty turned in 1903 with a request for a long-term contract for fuel oil at a time when naval strategists were eager for a secure source of oil to fuel the warships of the future. To this company the Admiralty turned again, a year later, to rescue William Knox D'Arcy's concession in Persia from the possibility of being lost to Britain, thus saving Persia itself from the risk of falling, perhaps irreversibly, into Russian hands. After many vicissitudes, Burmah Oil floated the Anglo-Persian Oil Company in 1909 as an all but fully owned subsidiary.

The Persian discovery proved so extensive that only a few years later the task of financing its rapid development, and in particular of making it the Royal Navy's prime source of fuel oil, led to the historic acquisition by the British government of a majority shareholding in Anglo-Persian. Having thus been the progenitor of what was to become the vast oil industry of the Middle East, Burmah Oil's own activities were to be confined for the next forty years or so to its operations in Burma and India.

Every oil company must to a greater or lesser extent parley with governments. Yet in the first century of its existence, Burmah Oil was fated over and over again to touch one or other of the sensitive nerve-ends of the British government. In the years 1915–24, the company found itself the kernel of ultimately abortive amalgamation talks aimed at creating an all-British combine that would have dominated Britain's oil scene between the wars, much as ICI was to do in chemicals.

Before the main story begins, however, we must trace the history of the period before 1874: that is, up to the year of the company's conception. As Chapter I will show, there had already been two decades of frustrated effort, and a further decade or so was to follow before the company's eventual birth.

CHAPTER I

The Forerunners

As its name suggests, the company was formed to extract, refine and market oil in Burma (known to the British until the latter years of the last century as Burmah). Although the Burmese had been working the deposits on a very modest scale for centuries, there is no precise early record of their activities.[1] However, a Chinese traveller is reported as having seen a thriving industry in operation towards the end of the thirteenth century.[2]

Predictably, legends have filled the vacuum left by history. The most celebrated legend describes a king of Burma who, in about AD 1100, is said to have used supernatural powers to turn some scented waters into malodorous oil during a royal progress round his dominions. This was after the ladies of his harem had played truant to bathe in those waters. For their disobedience, the ladies were summarily executed but at once avenged themselves by haunting the king with their spirits or "nats" – a weighty problem in that spirit-ridden country. He therefore appointed twenty-four guardians of the nats, and rewarded them with the ownership of the oil-bearing land.

That account and others have withered under the pitiless scrutiny of the scientists: at least one geologist found no evidence that the deposits could have been worked before about 1700. Yet, whatever their antecedents, the guardians' descendants – known as the Twinzas (or those who lived off the wells) – grew rich over the years on their smelly product. Like the oilfield, the Twinzas were real enough, although not exactly situated in a fairy-tale region of a fabulous kingdom: the setting

5

was no Shangri-la. On the contrary, it was a desolate spot, parched by the sun for most of the year and battered by torrential rainstorms during the brief monsoon season.

It lies in the centre of Burma's so-called dry belt, some 300 miles north of Rangoon and close to the Irrawaddy river. The locality bears the pleasant-sounding name of Yenangyaung, which translates less happily as the creek of stinking waters: it was there that the Twinzas followed time-honoured methods of hollowing out the wells and raising the crude oil.

Unschooled as they then were even in elementary technology, they had to dig the wells entirely by hand. The hard layers of rock they broke up by dropping heavy iron weights down the hole, and then hauling the weights and rock fragments to the surface in baskets. Placed over the well was a crossbeam, roughly fashioned out of a tree trunk in the middle of which was a groove; through that groove the rope was hauled by teams of coolies – usually women – running down an inclined plane to hoist the baskets to the surface.

The crude oil was found at a maximum depth of about 250 feet and the miners had to paddle knee-deep in the liquid which was hot, pungent and highly toxic. Until the 1890s, when they started to use a form of breathing apparatus, they could stay down for only 30 seconds at a time. They then needed half-an-hour's rest on the surface while others took their places. The wells never spouted; oil seeped into them slowly enough to require bailing out just once a day.

It was collected in locally made earthenware jars, many of which were smashed on the carts that carried them several miles along the heavily rutted and winding tracks to the Irrawaddy. By contrast, the ferrying on bamboo rafts down-river to Rangoon was a relatively smooth operation, enlivened only by the fast current and occasional sharp bends. At Rangoon, there was not even the crudest form of refining. The oil was sold in its raw state, for burning in the primitive lamps of the day and for proofing houses and boats against water and the ravages of insects; it was also popular as an embrocation, notably for rheumatism.

The first European traveller to provide a detailed description of Burma's primitive oil industry visited the area in 1795, when some seventy wells were producing an estimated total of 25,000 40-gallon barrels annually. He was an Englishman, Michael

Symes, sent on a mission to the kingdom of Burma by the Governor-General of India. The subsequent interest of the British in both the country and the oil, although for different reasons, needs to be explained. On the political side, Burma had a common frontier with Assam which later became India's easternmost province, and with the native state of Manipur; it also faced the Indian mainland across the Bay of Bengal. Symes's mission was primarily designed to counteract the allegedly over-strong French influence at the Burmese court. The French had withdrawn from India under the Treaty of Paris in 1763. Until 1885, however, Britain remained concerned that the French were seeking to establish their authority in Burma, perhaps so as to be able again to contest Britain's position in India.

It was, however, the King of Burma's own aggressive policies, culminating in his invasion of Manipur, Assam and other territories claimed by India, that led to the first Anglo-Burmese war of 1824–6. The British then annexed the coastal regions of Arakan and Tenasserim. As to the oil, it was only a year or two later that some samples of "Rangoon oil" came into the possession of George Swinton (d. 1854), secretary to the government of India in Calcutta. He forwarded these to the Royal Society of Edinburgh, whose council in 1830 invited (Sir) Robert Christison, professor of medicine at the university there, to analyse them.[3]

After considerable experimentation Christison was able, by distilling the crude oil, to take out some paraffin wax that was its most prominent substance. A colleague, William Gregory, the professor of chemistry, later went on to extract some liquid kerosene.[4] These two products, wrote Gregory in 1836, could be put to practical use once an easy and economical method of extracting them had been devised. Yet even the very clever and self-confident Victorian scientists, so boldly unravelling the secrets of nature, found that the transition from laboratory experiments under ideal conditions to effective refinery products was to be a painfully long-drawn-out process.

Not that scientists elsewhere in the world were having any greater practical success. In the United States Benjamin Silliman, Gregory's opposite number at Yale University, during 1833 made the first known distillation of mineral oil in the New World.[5] Yet his findings were not followed up until 1855 when his son Benjamin analysed some specimens from a seepage in Pennsylvania; the seepage was to lead four years later to Drake's oil discovery which, as described above, is generally regarded as the starting point of the present-day oil industry. There had been the earlier refining of coal-oil (or shale) on both sides of the Atlantic, most notably in Scotland by James Young[6] from 1850 onwards, but commercial refining of well-oil in America took place later than 1859 and therefore some years after Burmese crude oil had been successfully refined in Britain.[7]

The latter achievement owed much to the very gifted chemist and Fellow of the Royal Society, Warren de la Rue, a partner in the stationery and printing firm Thomas de la Rue & Sons.[8] As a juror for the Great Exhibition of 1851, he had come to know the directors of the largest candle-making firm in Britain, Price's Patent Candle Company Ltd of London. He sought to solve their problems over shortages of animal fats and vegetable oils, by developing processes for refining the Burmese oil that was so rich in paraffin wax: an ideal ingredient for candles. He patented his processes from 1853 onwards. These processes seemed technically so efficient that Price's directors devised a spectacular instance of Victorian endeavour, one that has been almost entirely overlooked by historians.

The directors had been helped by the outcome of the second Anglo-Burmese war of 1852, sparked off by provocative actions on the part of the king's officials against British merchants, which the British authorities felt they could not ignore. Britain thereafter annexed Lower Burma, including Rangoon which rapidly developed into a prosperous seaport and trading centre. It was a new King of Burma, Mindon Min (1853–78), in the now landlocked kingdom of Upper Burma, or Ava, who proclaimed oil to be a royal monopoly and required the Twinzas to sell their whole output to him for Rs1 ½ (10p) a barrel. As the oil was resold to customers for at least ten times that figure, he acquired considerable revenues, although the Twinzas benefited as well by enjoying a guaranteed market for their output as well as certain tax exemptions. Price's directors were therefore able to arrange through agents they sent out specially to purchase crude oil, transport it to Britain, and refine it in their factory at Battersea. While making use of the wax themselves, they planned to sell the by-products to other customers.

Yet the consequences of this initiative were to carry Price's to the brink of disaster and have serious repercussions in Burma itself. The agents contracted for crude oil to be shipped from Rangoon at the rate of nearly 2,000 barrels a month. This was ambitious, considering that in 1855 there were probably no more than 130 active wells in Burma, yielding about 46,000 barrels a year, or nearly double the output of 1795. On reaching Rangoon, the oil was poured from the jars into a variety of containers, mainly large metal tanks of 8 or 64 cubic feet capacity specially sent out from Britain. Ships chartered by such London importers as Frith Sands & Co. of Old Broad Street and William Syers & Co. of Rood Lane, then conveyed this inflammable substance, as part of mixed cargoes, under sail on the 12,000-mile voyage, often of five to six months duration, round the Cape of Good Hope.

In June 1857, consignments began to arrive in quantity at Price's Battersea factory. Before the end of the year 18,400 barrels had been entered through British customs,[9] and an even larger quantity of 21,200 barrels arrived in 1858. With such shipments coming in, the company seemed to be more than achieving its objectives. Then it ran into crippling difficulties.

The operation's success depended on selling the maximum amount of refined products at minimum cost. Yet freight and associated charges accounted for a third of the £6.00–£7.00 per barrel that the crude oil was worth in London. As for output, refining at Battersea was by later standards inexact and wasteful and (using de la Rue's experimental percentages as a rough guide), the yield of paraffin wax, as the most sought-after product, was only 10 per cent. In any case, by that time it was no longer the revenue-earner originally expected, as candle prices generally had been badly hit by competition.

Nor did the company have the success it had hoped for over the by-products. Lubricating oil, comprising 40 per cent of the total yield, should have been a money-spinner; yet mill owners in the north of England were reluctant to use that comparatively untried brand in their machinery, despite its cheapness compared with existing non-mineral lubricants such as sperm-oil. The very light oils (20 per cent) were not of much value except as solvents and detergents. Even kerosene (20 per cent) then had little sale among British customers. Outside the urban areas fed with town gas, the wealthy used expensive vegetable-

or whale-oil in their lamps and the rest continued to burn candles. In those same years James Young was laboriously building up a market for his shale-oil kerosene, selling a special lamp for burning it. Price's, without his marketing skills, had to dispose of its entire kerosene output to certain American manufacturers of coal-oil.

Once those disappointing results had become apparent in March 1858, the company came to a hard decision: it cancelled any further deliveries of crude from Burma. However, many months were to elapse before the final shipments arrived, leaving it with considerable stocks of crude oil and various refined products which it did not succeed in liquidating for the best part of a decade.

All these events occurred well before Drake's oil discovery of 1859 in Pennsylvania. The earliest consignments of United States crude did not in fact reach England until the end of 1861, and refined products later still. Thus Price's directors were wrong to suggest, as they did later, that Burmese shipments had to be halted because of competition from America.[10] Yet it was American and later Russian kerosene which thereafter satisfied the soaring market in Britain, with transport costs far below those from the east.

So ended one of the bolder commercial ventures of the early Victorian era. It involved annual expenditure of more than £120,000 on crude oil alone, but the suspicion remains that it never had any real chance of success. As shown above, transport costs were particularly onerous; yet had the scheme not run into trouble so soon, Price's might have erected a distillation plant at Rangoon to reduce costs, just as they had in the 1840s installed coconut-crushing machinery in Ceylon so that the less bulky vegetable-oil rather than the coconuts themselves could be exported to Britain. Other causes of the débâcle, such as difficulties over marketing the main product and by-products profitably, should perhaps have been foreseen.

In Burma, much confusion was caused by this quite unexpected turn of events.[11] A consortium of East India merchants at Rangoon had helped to finance the operation, and now sought to retrieve matters by arranging to refine on the spot and trying to sell what they could in Burma and in markets nearby. As early as 1858 an oil works was in existence at Dunneedaw, near the spit of land at the confluence of the Rangoon and Pegu rivers.[12] The exact status of the proprietor is not known, but by 1861 he had disappeared when the merchants acquired on lease a 30-acre site, no doubt also at Dunneedaw, and erected some stills there.

By the following year, a distillery of sorts was completed, but failed, because of technical snags, to come on stream for many months. To be sure, crude oil had never before been refined in the heat of the tropics; apart from Price's operation, about the only earlier refining on any scale was of some well-oil in Russia and the United States, and shale-oil in the United States and Britain.

Later in 1863, the Rangoon proprietor died and the refinery passed to Morris Roberts Syers, connected with the William Syers & Co. which had imported oil on Price's behalf.[13] His interests lay elsewhere. He decided to sell out in order to establish – of all things – a palace of entertainment in London. In October 1864, he opened the Strand Music Hall and later took over the Oxford Music Hall in Oxford Street; he thus provides what must be a rare early link between the oil business and show business.

The purchasers of the refinery at once formed the Burmese Oil Distillery Ltd in April 1864.[14] Those involved in its formation were mostly East India merchants based in London, one of whom was David Wilson. He was a brother of the joint managing directors of Price's and had for many years previously been the local manager of the company's operations in Ceylon. He had lately returned to England to become a full-time company promoter.

The new company's capital was £100,000, of which £45,000 was immediately offered to the public. The prospectus blamed the considerable losses incurred by previous importers into England on the fact that only half the crude – presumably the wax and lubricants – had become saleable products. New capital would allow a more efficient refinery to be built in Burma, so that transport charges would be minimized and better quality products would find a ready market in the east.

The chairman, John Ogle, an East India merchant, was early involved in an unfortunate brush with *The Times*, which on its business page criticized a passage in the prospectus stating that the new American duty on Pennsylvanian oil would in

effect give Burma a direct bonus of about 15s (75p) a barrel: *The
Times* pointed out that the duty was not levied on American
exports. Not until the paper had repeated the charge did he
grudgingly admit his mistake. Even his counterassertion that
all the kerosene consumed in India and Australia was equivalent to twice the distillery's output cut little ice since there was
no reason to expect those countries to buy from the company.

Still more damagingly, the *Rangoon Times* reported that the
Burmese Oil Distillery was paying £20,000 for the plant and the
lease of the land, although those assets were worth only a third
of this sum. "At this rate," the paper commented sarcastically,
it "will get on like a crab, whose progress forwards is attained
by going backwards." Yet for the next year or so the company
seemed to be progressing in the right direction as the refinery,
with a stated capacity of 34,000 barrels a year, began to turn out
apparently worthwhile quantities of the various fractions.[15]

At first the company made some exports to France, 4,000
barrels a year in both 1864/5 and 1865/6, or about one-eighth of
total production. No doubt these were arranged through the
agent it had specially appointed in Paris; hardly welcome news
to government of India officials, constantly worried as they
were over France's possible designs on Upper Burma. French
and other continental refiners had earlier bid for the whole oil
concession from the king's agents, but the Rangoon merchants
had beaten off that challenge.[16] Then early in 1865 a French
adventurer, the self-styled "General" D'Orgoni, already a
source of lively irritation to the Governor-General's staff in
Calcutta, secured the oil concession as from that October, only
to die before the agreement was due to come into effect.[17] With
his death, the French connection ceased in regard to oil, but
without allaying official suspicion.

All exports of Burmese oil to European countries dried up in
1866 since they could not compete with cheaper American oil.
Only some insignificant shipments to mainland India and the
Straits Settlements (Malaya) continued. Investors had never
fully shared the prospectus's sanguine views, taking up only a
minority of the shares on offer, so that much-needed cash had to
be raised by 7½ per cent debentures. Faced with inadequate
sales, in 1867 the company had to renegotiate downwards its
contract with the king, who had insisted on maintaining the
required deliveries of crude despite political disturbances set off

by a palace revolution which nearly unseated him. Then in the autumn of 1868, it could not redeem or renew the debentures coming due for repayment, and steps had to be taken to wind it up voluntarily. Even so, it struggled on for a few more years, helped by cash called up from increasingly reluctant investors.

The new company to take over was Scottish: the Rangoon Oil Company Ltd, registered in Edinburgh in July 1871.[18] The promoter, chairman and managing director was James Galbraith (1818–85), a shipping magnate who also headed the well-known Glasgow shipowners, P. Henderson & Co., famous to generations of travellers as Paddy Henderson.[19]. His energy, prevailing over chronic bad health, had carried him from humble origins to a leading position in the city's maritime affairs. For nearly a decade his fleet, returning almost empty on the New Zealand run, had called in at Rangoon for return cargoes, mainly rice and teak. But as recently as 1870 he had begun to convert that service to steam, using the newly opened Suez Canal and thereby saving on both time and costs.

Together with T. D. Findlay, a fellow Glaswegian and owner of the mercantile firm, Todd Findlay & Co. of Rangoon, he had in 1865 established the Irrawaddy Flotilla Company to run a river transport service, immortalized by Rudyard Kipling in his poem "Mandalay" as the "old Flotilla", with its "paddles chunkin' from Rangoon to Mandalay". Now that he had an interlocking water transport system that ran past the oilfields close to the Irrawaddy, via Rangoon to Britain he was, according to an official source, anxious to re-establish exports to Europe. There he was oversanguine, given the flood of American imports now crossing the Atlantic; what he did do instead was to build up shipments to mainland India to an average of nearly 11,000 barrels a year or five times the average in the period 1864–71. He thereby established India as the natural overseas market for Burmese oil products. The investment he put into the refinery at Dunneedaw was long remembered as the earliest attempt at refining in Burma on a truly scientific basis.

He did adopt one noteworthy improvement in wax making, which involved sweating it out rather than squeezing it under pressure; that process had been evolved by Price's in London. Kerosene, however, remained smoky and discoloured. Yet

Indian customers were willing to buy whatever could be shipped there, as it was better than the vegetable substitutes and far cheaper than imported kerosene from the United States.

What broke the Rangoon Oil Company was not the failure to rebuild a market in Europe but the arbitrary actions of the king's government in Upper Burma. Until 1874 the monopoly agent for oil had been the Woon or governor of Yenangyaung, who farmed the revenue for a fixed annual sum. He had, to be sure, practised some forms of discrimination, such as charging different prices to buyers in different areas, according to the pressure of demand.

Then in 1874 there emerged a new monopoly agent, Moolla Ismail, a merchant from Surat in India and a long-time resident at the king's court. He went further than his predecessor by having the Yenangyaung oil, abundant in wax, mixed with the oil from Yenangyat, some 50 miles further north along the Irrawaddy where indigenous production had started as recently as 1864. The latter oil was far less popular since it contained more light fractions, then of small commercial value. He also held up shipments from time to time: deliveries to Rangoon in 1875/6 and in 1878/9 were little more than half the levels of previous years.[20]

Later evidence makes clear that the purchasing firms had to give financial inducements to both Moolla Ismail and certain ministers to ensure even a minimum supply. Official memoranda after 1886, for instance, mention "the constant bribes and . . . other extortions" to which the Rangoon Oil Company had been "undoubtedly exposed in the late king's time".[21] An impartial American historian has described that regime as having been "an autocracy tempered by bribery",[22] The tempering process was plainly calamitous for the company. Galbraith spent in all about £30,000 on his new refinery; yet as long as extortions and uncertain oil deliveries persisted, he had little prospect of making the company pay. In mid-1876 he put it into voluntary liquidation.

On three separate occasions in the preceding thirteen years, then, successive British promoters had failed in their efforts to purchase, refine and market Burmese oil and achieve adequate returns. To many observers the latest collapse must have seemed like the end of the road for British enterprise there, at least for a long time to come. Indeed, it might well have turned out to be so had not a man come forward to acquire the moribund company's assets. He was a Scottish-born East India merchant named David Cargill.

Notes

1 For a full account of the legends, successive descriptions of the oilfields from 1795 onwards and scientific evidence, as well as information on the products and their marketing, see F. Noetling, *The Occurrence of Petroleum in Burma and Its Technical Exploitation* (Memoirs of the Geological Survey of India, XXVII, Part 2, 1898). This indispensable source of data to 1895 is ably continued by Pascoe's work; see note 2 below.

2 E. H. Pascoe, *The Oil-Fields of Burma* (Memoirs of the Geological Survey of India, XL, Part 1, 1912), p. 74n refers to the discovery by a French archaeologist in the library at Peking of the Chinese traveller's diary, but gives no more definite information. The distinguished Sinologists the late Arthur Waley and Dr Joseph Needham were unfortunately unable to throw further light on this intriguing reference.

3 As Christison's and later experiments are so well documented in Noetling's account (see note 1 above), pp. 143 ff., references in notes below are given only when supplementary information needs to be provided.

4 The spelling "kerosene" is used throughout the present book since it "remains the usual one in general usage and still occurs in technical contexts", R. W. Burchfield (ed.), *A Supplement to the Oxford English Dictionary II H–N* (1976) p. 486. "Kerosine" was not adopted by the (British) Institute of Petroleum until after 1924 and the older spelling is retained in many oil histories published since then.

5 The 1833 distillation is mentioned in H. F. Williamson and A. R. Daum, *The American Petroleum Industry I: The Age of Illumination 1859–1899* (1959), p. 68, which also summarizes the 1855 analysis on pp. 69–72.

6 James Young's contribution to oil technology is at last being placed on a historical basis by J. Butt, "Legends of the Coal-Oil Industry 1847–64", *Explorations in Entrepreneurial History*, 2nd series II (1964–5), pp. 16–30, and by O. Colverd, *Institute of Petroleum Review*, 14 October 1960, p. 343, ibid., 15 March 1961, pp. 65–7, and 16 March 1962, pp. 84–6, and "Scottish Branch" ibid., 15 September 1961, pp. 295–8.

7 Basically there were then two distinct processes of "refining", namely (i) distillation: subjecting the crude oil to steadily increasing heat so as to vaporize it, and then liquefying the various fractions by condensation, and (ii) refining: purifying some of these fractions, usually by chemical treatment.

8 Regrettably, no biography appears to exist for W. de la Rue. His article, with Hugo Muller, "Chemical Examination of Burmese Naphtha, or Rangoon Tar", is in *Proceedings of Royal Society*, VIII (1856), pp. 221 ff. The patents he took out for treating Burmese crude are no. 1748 of 1853, nos 1051 and 2719 of 1854, and no. 2002 of 1855. His connection with Price's is referred to briefly in (Price's), *Still the Candle Burns* (privately printed 1947) pp. 32–3, and more fully in G. Wilson, *The Old Days of Price's Patent*

16   *A History of the Burmah Oil Company*

- *Candle Company* (privately printed 1876) and Price's, *A Brief History of Price's Patent Candle Co. Ltd.* (privately printed 1891).

9 Information from Customs and Excise Records, by permission of the Board of Customs and Excise.

10 Price's, *A Brief History of Price's Patent Candle Co. Ltd.*, pp. 10–11.

11 Valuable information is in the pioneering article by O. Colverd, "The Early Days of Burmese Oil", *Petroleum Review*, 26, 1972, pp. 237–41.

12 A. G. Roussac (compiler), *New Calcutta Directory*, 1858, p. 203. The proprietor was James J. Wallis and the assistant George Bristow.

13 For M. R. Syers see F. Boase, *Modern English Biography 1851–1900*, VI, (1921), col. 653.

14 The company file is in Public Record Office (PRO), BT 31 939/1214C. Prospectus in *The Times*, 25 April 1864, and article on business page on same day. See also further article 28 April and letter from Ogle, 29 April.

15 *The Grocer*, IX, (Oil Trade Review), 5 May 1866, p. 38.

16 Williamson and Daum, *American Petroleum Industry*, I, p. 323.

17 Oliver B. Pollak, *Empires in Collision: Anglo-Burmese Relations in the Mid-Nineteenth Century* (1979), p. 125.

18 File in Scottish Record Office, BT 2 361.

19 For Galbraith see *Portraits of 100 Glasgow Men* (c.1890), I, pp. 141–4, and D. Laird, *Paddy Henderson* (1961).

20 Data and commentary on deliveries from Upper Burma and shipments from Rangoon are in *Reports on Trade and Navigation of British Burma* 1874/5 onwards (*of Burma* 1885/6 onwards).

21 India Office Library (IOL) R&A (Minerals) Proceedings 1889. Secretary for Upper Burma to Chief Commissioner, December 1886.

22 J. LeRoy Christian, *Burma and the Japanese Invader* (1945), p. 356.

CHAPTER II

David Cargill and Kirkman Finlay
1874–86

Although the Burmah Oil Company was not born until 1886, the concept of its creation can be dated precisely: from the closing months of 1874, with the initial encounter of two men who were subsequently to nurse the none too robust infant into healthy adolescence. They met at the Rangoon office of Galbraith Dalziel & Co., managing agents of the Rangoon Oil Company. The office was little more than a timber shack, on top of a single-storey brick house that stored the agency's wares, ranging from piece goods to imported luxuries. In some ways those wares were better accommodated than was the office staff, who long remembered the timber erection as "villainously hot" during the seemingly interminable summer, although tolerable enough in the cool winter season.

The two men who were to dominate the company's formative years into the new century were both Scotsmen. They were also younger sons in their respective families: the ones, according to the Gaelic proverb, who had to take up the book, the oar or the sword – although by the 1870s alternative possibilities existed which could have been more appropriately symbolized by the clerk's pen or engineer's oily rag.

The senior of the two, David Cargill (1826–1904), was an east coast man from Arbroath. He had thirty years' experience of trading with Ceylon, first in Colombo with the import-export firm of William Milne & Co. After Milne's death in 1861, he had returned home to acquire the business and run it from its Glasgow head office.[1] His life had been busy enough, if rather lacking in incident. Then, from the beginning of the decade, he found himself abruptly jolted out of his accustomed routine. The Ceylon business, now known as Cargill & Co., had become highly lucrative, being later dubbed "the Harrods of the East"; but his other commercial venture in the island – coffee growing – had of late been stricken with a disease that was soon to wipe it out.

Two bereavements had followed in quick succession: his father's death which brought him a legacy, and in 1872 the death of his wife, leaving him with five small children. On its formation in 1871, he had almost casually taken up a few shares in the Rangoon Oil Company, because many of his fellow-members of the Glasgow Chamber of Commerce and Merchants House had done likewise. A year or two later he reluctantly agreed to become a director. When in 1874 he planned a trip to visit his properties in Ceylon and convert them to what soon became the island's staple product, tea, he was persuaded to go on to Rangoon and inspect the oil company's agency there. That was how he came to meet Kirkman Finlay (1847–1903), the agency's senior assistant, who was in effective charge.

The younger man bore a respected Glasgow name, for a Kirkman Finlay (1772–1842) had been Lord Provost of the city as well as its MP and, as a leading business man there, an acknowledged promoter of its nineteenth-century commercial greatness. Kirkman Finlay the younger was in fact the second son of a jeweller who, in his affluent days, had numbered the Lord Provost among his friends and named his son after him, but he had later fallen on hard times before dying prematurely. The younger Finlay proved hardy enough, having been brought up at Brodick Castle on the island of Arran, where his grandfather had been head gamekeeper to the Duke of Hamilton.

His career, like Cargill's, was moving through sheer chance towards the founding and direction of a major oil enterprise. He also had sought his fortune overseas, initially in Angola, before joining the agency. But he happened to be a brother-in-law of the younger James Galbraith (1834–1919) who, although not related to his namesake (see Chapter I), was responsible for the Rangoon Oil Company's agency in the east. Galbraith, then constantly on the move between Glasgow and the east, arranged for his brother, who had married another Finlay sister, to assist in the Rangoon office, but that brother had died unexpectedly. Finlay was invited to take his place, and arrived at Rangoon in 1869, very soon afterwards travelling up the Irrawaddy to see the oilfields. He went on to the royal capital of Mandalay, where he met – or at least set eyes on – Mindon Min.

While in Burma, Galbraith strove hard to discover oil in British territory and thereby break the king's monopoly. He had received from the chief commissioner for British Burma, Sir Ashley Eden, a "roving commission" to prospect for oil, and a promise of exclusive development rights in the event of success.[2] Thus he was glad to have a man of Finlay's abilities to carry out the day-to-day work of the agency. His partner, Alexander Dalziel, seems to have been an alcoholic who later died of apoplexy; according to an authority on the period, "medical reports of the 1870s and 1880s show that a surprisingly large percentage of Europeans in the east died from the effects of over-drinking."[3]

The two men who now met were of contrasting appearance, Finlay having a weak face with hair parted in the middle and a long-ended moustache in the Chinese manner. Cargill's features, on the other hand, were striking enough, and he heightened them by brushing his hair well back to emphasize his high forehead. If his eyes were small and had a hint of meanness, a Roman nose dominated his face and a full beard all but covered the broad knot of his tie. So closely did he resemble a seafaring man that he and Finlay might have been enacting a scene from one of Joseph Conrad's later novels: the sea-captain, newly docked in the east, paying a business call on the local agent.

Yet their characters were usefully complementary. Cargill was unemotional, tenacious in pursuit of his objectives and at times careless of others' feelings, even those of his own family. Finlay, on the other hand, had a warm interest in people and preferred to succeed through intuition and persuasion rather than crude bargaining power. What exactly they talked about is not known, but Cargill's son later spoke of Finlay's "implicit belief" in a great future for oil in Burma, and of the insistent advice he gave Cargill to help secure that future.[4]

Even the enthusiastic Finlay could scarcely have believed that Cargill would heed his advice, for Cargill did not visit the Burmese oilfields and in fact never went out east again. He was by then approaching 50, the kind of age when a man of his bent begins to dream of an easier life; he therefore contented himself with a thorough inspection of the refinery. Gradually, the project took a hold on his imagination. Once back in Glasgow, he began some detailed consultations; then in 1876 he took his momentous decision. He offered £15,000 for the Rangoon Oil Company, just half of what had been spent on the refinery alone since 1871. Not surprisingly, his colleagues were only too anxious to cut their losses. They accepted his offer with alacrity and had the company wound up. Thereby he condemned himself, for the next ten years or so, to constant anxiety and financial stress.

Continuing to use the Rangoon Oil Company's name and goodwill, Cargill established a fresh and more effective management structure. As the younger Galbraith had already gone

The Burmah Oil Company's first office, c.1890, at 191 West George Street, Glasgow.

bankrupt, Cargill was able to set up an entirely new agency in Glasgow, called Milne & Co., which operated alongside his own business of William Milne & Co. at 191 West George Street. That was a modest three-storey building which must have become rather cramped as the interlocking businesses grew in size and complexity.

Finlay gladly set up an agency in Rangoon, named Finlay Fleming & Co., in which Milne & Co. was the majority shareholder in association with Finlay and a newcomer, Matthew Tarbett Fleming, who had until then been a Rangoon employee of the Glasgow-based British India Steam Navigation Company Ltd. The latter hailed from the higher ranks of Glasgow society. His father, John Gibson Fleming, had been a celebrated physician there who died in 1879, leaving his son a third share in a £60,000 estate. The younger Fleming was therefore able to bring some much needed capital into the agency.

Since the late 1860s, the electric telegraph had connected Britain with Rangoon and also, on the king's orders, with Mandalay. Thus Cargill had a means, which his predecessors had neglected, of learning very quickly about events in Upper Burma at a time when letters took as long as five or six weeks in either direction. In October 1878 he was able to write to *The Times*, quoting a cable just received, to the effect that King Thibaw had succeeded Mindon Min. He thereby scooped the story which appeared in the paper's news columns only 24 hours later. "Quietness prevails," Cargill informed *The Times*'s readers.[5]

Just as Cargill was the first to inform the British public that Thibaw was the new Burmese king, so Finlay was the first European that monarch received in audience, at Mandalay a few months later. Finlay brought with him a letter of introduction to the Foreign Minister, the Kinwun Mingyi, who duly arranged the audience with the king. He and the friend who accompanied him, had to remove their boots – or rather, have them removed by servants – and bow deeply in the Burmese manner: gestures of respect which British officials had for some years refused to make and which, absurdly enough, had precipitated a breakdown of relations between the two countries. Since history has bestowed on Thibaw a reputation for spinelessness, it is worth noting that Finlay, in his private diary of the visit completed within days of returning to Rangoon, noted his "bright penetrating eyes and well set mouth" and the impression he gave of "intelligence and determination".

At the same time, Thibaw seemed "uneasy and nervous" as well he might, since only a week or two later he gave in to the demands of his strong-willed consort and had some eighty members of the royal family massacred to eliminate potential rivals. The next European to visit him, a German doctor, also formed a sympathetic first impression, but with hindsight saw indolence and cruelty in Thibaw's handsome features.[6] Yet most observers, including Finlay, fervently hoped that this initial blood-letting, however much to be deplored, would give way to a period of political stability and of greatly improved Anglo–Burmese relations.

During 1879, Finlay returned to Britain and became Cargill's righthand man in the Glasgow office. Fleming also went home soon afterwards. They appointed as general manager the Burmese-speaking James Fisher, who maintained liaison with the court at Mandalay through a fellow-employee of Persian extraction, Aga Mohamed Ameen. When Fisher died in 1886, he was succeeded by a more plodding man, Alexander Pennycuick.

If good organization alone could have brought profit to an enterprise, then Cargill would have found himself becoming a rich man. In fact, during the period 1876–89 he earned little or no money from Burmese oil and had to dip into his own pocket to pay for the investment and prospecting work required. By 1889 he had spent £100,000 on his oil activities; net expenses, over and above receipts from sales, were thus running at about £6,000 a year. Although he could afford to meet these sums out of his income from general trading in Ceylon and Burma, that could not be expected to support his oil losses indefinitely.

His main problem overseas was the monopoly agent, Moolla Ismail, who continued to make life difficult for him as a purchaser of crude. Fisher's efforts must have been of some help, but not decisively. Perhaps oil consumers of the 1970s and 1980s, also much exposed to monopolist producers, could better understand the problem in Burma a century before than could the intervening generations. Moolla Ismail had apparently abandoned the old long-term contracts and insisted on spot transactions, or separate bidding for every cargo of oil required. Finlay Fleming & Co. later explained to civil servants in Rangoon how, in those years, it had sought to buy most of the Twinzas' output of oil and had "either to pay the monopolist very much what he demanded or else to close the works".[7] That kept the cost of crude oil onerously high.

The degree of monopolistic pressure can be illustrated by figures of oil deliveries from Upper Burma between 1875/6 and 1883/4, which were down 20 per cent on average compared with the previous nine years from 1866/7 onwards. Moreover, fluctuations from year to year were greater than previously and bore no relation to fluctuations in the Twinzas' output. Moolla Ismail claimed that the Rs250,000 (over £16,500) a year he paid the court for the oil monopoly rights exceeded the income he received so that he allegedly had to make up the deficit from the monopolies over customs and river dues. Yet the evidence suggests that each oil delivery had to be wrung out of him by substantial bribes.

Like James Galbraith, Cargill tried hard but unsuccessfully to by-pass the monopolist by discovering crude oil in British Burma. While in Rangoon during 1874–5 he had interviewed Sir Ashley Eden, who had earlier been so encouraging to Galbraith. Eden promised Cargill official assistance over crude supplies to ensure that the refinery was kept operating. Presumably Eden had thought he could exert British diplomatic pressure on the court at Mandalay: an impossible task, by the early 1880s, as continued political tension prevented any British representative from being stationed in the capital.

Cargill arranged to prospect at Padaukpin, a few miles from Galbraith's abortive operations at Thayetmyo. Although he spent the equivalent of £4,500 in tests, very little oil was raised. Yet almost against all reason, he remained optimistic. In 1882, he bought the freehold of the refinery site and spent much money in enlarging and improving the works. Here is one of the mysteries of entrepreneurship. Had he ultimately failed, his fellow-merchants in Glasgow would have branded him as utterly reckless and pig-headed. Indeed even after 1882, the quality of his kerosene was not high since distilling remained inexact. No wonder American imports of superior kerosene began to gain a foothold in Burma, aided by the abolition of the 5 per cent import duty in 1878. Quantities rose from virtually nil in the mid-1870s to 90,000 barrels in 1884/5, when deliveries from the Rangoon refinery were only 25,000 barrels and had to be sold at half the American price.

These setbacks were the more depressing because certain other oil prospectors in British Burma seemed to be doing rather well. The Rangoon government granted them concessions in the islands of Akyab where, early in the 1880s, they drilled down to 1,000 feet; the first time mechanical drilling took place in the province. When the wells began to flow, the respective operators set up two companies for the purpose, the Barangah Oil Refining Company Ltd and the Arakan Petroleum Company. Their joint forecast was that they would soon be producing at twice the rate then being achieved by the Twinzas. Soon, however, they began to suffer the fate that had attended so many other oil ventures around the world: production declined disastrously and led to one of those companies being wound up in 1885.

By that year, things appeared at long last to be looking up for David Cargill. In 1885 he obtained oil deliveries from Upper Burma that were the highest since the mid-1870s, at a price nearly a third lower than in the previous year. However, with conditions worsening daily at Thibaw's court, Cargill was soon confronted with some very unwelcome news. The oil monopoly, after twelve years in Moolla Ismail's hands, was once again to be put up for auction at Mandalay. According to strong rumours, the Moolla and his uncle would be bidding against one another, and whoever won was bound to exact even higher rates.[8]

Yet Cargill refused to cut his losses. Instead, in July 1885, he wrote to Lord Randolph Churchill, newly appointed as Secretary of State for India, asking how he intended to deal with the deteriorating conditions in Upper Burma.[9] The formal acknowledgment he received was perhaps no more than was to be expected. Of everyone concerned on the British side, he was possibly the most hopeful that things could work out to his satisfaction.

What were the grounds of Cargill's reluctance to give up? He had more than ten years of information-gathering on Burmese affairs behind him and, as his letter of July 1885 made clear, he had repeatedly drawn the attention of successive ministers of the crown, and the Rangoon authorities, to the consequences of the imminent collapse of Thibaw's kingdom. That letter also referred to a plea for outright annexation or the establishment of a British protectorate, in a petition signed by a number of prominent commercial men in Rangoon, no doubt including Finlay Fleming & Co.'s general manager. Cargill was in effect saying that as the kingdom was falling apart, it was about time that the British went in to safeguard the oilfields among other assets.

Only a few months later later, in November, the third Anglo–Burmese war of 1885 erupted; its causes have been fiercely debated ever since. The opinion of some historians, that Britain was hell-bent on dismembering the kingdom out of commercial greed, has in recent times given way to a more relaxed view, notably among American historians who have made the subject their own: some have even expressed surprise that Britain tolerated the tiresome regime for so long.[10]

An interesting study, based mainly on Burmese sources, has attributed the growing tension of 1883–5 to a single underlying cause: a drought in the dry belt where the oilfields lay, which was brought on by the wholesale destruction of the teak forests. That deforestation is said to have removed the capacity of the soil to retain the winter rainfall, leading to increased evaporation. The resulting revenue losses exacerbated the political instability of Thibaw's court, encouraged certain rebels in occupation of northern parts of the kingdom, fostered French adventurism and alarmed British officials and commercial interests. Against this background, 250,000 people are thought to have emigrated in 1884 to the prosperous tranquillity of British Burma, and thousands of others would have followed but for brutally repressive measures ordered by the king.[11]

In these circumstances, the threat from France became far more than the trumped-up excuse it was once claimed to be. The French were currently consolidating their hold on the north of Indo-China, which had a common frontier with Burma only 300 miles from Mandalay, the approaches to which crossed the Shan States that were already in insurrection. Even more significantly, the pretender to the Burmese throne was in French hands. Anglo–French rivalry in south-east Asia, to quote a revered British historian who was no great friend of imperialism, was serious enough even to represent a potential cause of a European war.[12] Another British historian, in the confidence of certain top men of the time, judged that France, Germany and Russia were, in 1883–5, making a "dead-set" against British interests in various parts of the world. This move had the unanticipated consequence that it "awakened British statesmen from their apathy, and led them to adopt measures of unwonted vigour".[13]

Indeed, in some ways the great powers' rivalry paralleled that over Persia two decades later. In Persia also, the rule of the central government was falling apart, and the consequent power vacuum invited the prospect of intervention by a country, Russia, which could well threaten Britain's hold on India. As Chapter VII will show, there was one piquant difference in these two cases. Whereas in 1885, Cargill and his colleagues were passive and impotent bystanders, by 1904 the company he had formed, and his son as chairman, were to play a central part in combating the possible threat of Russia's economic imperialism in Persia.

In Upper Burma, the bankrupt Thibaw's efforts to establish relationships with France and Italy as counterweights to British pressure appeared highly suspect to Britain. In mid-1885, a French consul, F. Haas, arrived at Mandalay and obtained a vital railway concession, offering enormous loans partly secured on the oil revenues. When shortly afterwards the directors of the Bombay Burmah Trading Corporation, which dealt in another monopoly product – namely teak – received a penal fine for alleged irregularities over the payments of dues, many believed with some justification, according to reliable accounts, that Haas had encouraged the king's hopes of Bombay Burmah's teak concession being expropriated and taken over by more pliable French interests. The episode of the fine, which actually precipitated a British ultimatum and the despatch of the task force to conquer Upper Burma, was merely the occasion and not the cause of the war. The king brought about his own downfall, not so much by refusing to budge over the fine as by replying defiantly that "friendly relations with France, Italy and other states have been, are being and will be maintained."[14]

Throughout that summer and autumn, Cargill closely followed every move from Glasgow, exchanging cables and longer messages by post with Rangoon. Ten days after the despatch of the ultimatum, at the end of October, he instructed Finlay Fleming & Co. to request the government of British Burma for concessions of about 4 square miles at Yenangyaung and Yenangyat, without infringing "the just rights of the Twinzas". His request was a shrewd pre-emptive move and he received the remarkably stiff and convoluted reply that "the matter would have attention in the event of circumstances turning out such as would admit of any such arrangements being given effect to."[15]

Circumstances did, however, begin to turn out that way when the *Glasgow Herald* reported on 10 November: "War has been declared against King Thibaw. No prospects of business this month. Prospects are good afterwards."[16] Those words neatly summarize British commercial attitudes to a whole era of colonial wars, and perhaps Cargill's feelings after a decade of frustration and financial loss. By the end of the month, Upper Burma was overrun and Thibaw swept from his throne.

The upshot of that conquest came up to the expectations of remarkably few on either side. Any chance of replacing Thibaw with a more acceptable Burmese monarch proved impossible, not least because the pretender was still in French hands. Instead, much to the disgust of Rangoon's commercial men who had hoped for a separate crown colony or protectorate to be established, the new unified province of Burma was to become part of the Indian empire. Not only would Burmese interests thereby be subordinated to Indian ones, but the revenue surplus derived from the province's thriving commerce and industries would be drained away into the Indian treasury.

There was disappointment over oil matters as well. The new British administration in Burma, seeking diligently for the minerals that would strengthen the province's economy, "went nap" on coal, as one knowledgeable civil servant of the day put it.[17] This proved uniformly disappointing because the coal, unlike the versatile oil, was of very poor quality. In their turn Cargill and Finlay, although free of the king's and his agents' capriciousness, were to find British officials doctrinaire and obstructive, being concerned more to prevent the creation of an oil monopoly than to foster the development of a really efficient industry, even under suitable safeguards. The Burmah Oil Company was never to achieve an easy relationship with government; even so, after 1886, the opportunities at last existed for carrying out really worthwhile oil operations in Burma.

Notes

1 "The Origin of the Big Oil Companies", *Glasgow Herald*, 8 August 1944, and letter by W. D. Cargill Thompson, *Glasgow Herald*, 12 August 1959.

2 IOL P 698 R&A Proceedings, (Mineral & Geological Survey) 1871–5, especially November 1871, August 1872 and May 1873.

3 N. Sherry, *Conrad's Eastern World* (1966), p. 106.

4 1937 AGM, *Glasgow Herald*, 5 June 1937.

5 *The Times*, 2 October 1878; Mindon Min's death reported 3 October.

6 A. T. Q. Stewart, *The Pagoda War* (1972), p. 62.

7 IOL P 2729 R&A Proceedings, August 1886; memo of Chief Commissioner on Yenangyaung, 1 July 1886.

8 ibid.

9 Command Paper C 4614 (1886), "Correspondence Relating to Burmah", no. 96. Milne & Co., Glasgow to Lord R. Churchill, 6 July 1885, p. 166.

10 The earlier view is probably best exemplified in D. Woodman, *The Making of Burma* (1962) ch. XI. Maung Htin Aung, too, in *The Stricken Peacock: Anglo–Burmese Relations 1752–1948* (1965), regards British commercial interests as the main culprits, but in *A History of Burma* (1967), p. 257, he does acknowledge British suspicions of French aggressive and ambitious policies in south-east Asia. Did these suspicions have any foundation? Stewart, *The Pagoda War* (especially p. 72) seems to think so, while D. J. M. Tate, *The Making of Modern South-East Asia I The European Conquest* (1971), p. 418 considers that there was probably French intrigue behind the episode of the Bombay Burmah Trading Corporation. Britain's undue toleration of her "uncongenial neighbour" is in Pollak, *Empires in Collision*, p. 184. See also notes 12 and 13 below. Events in London, where the hawkish Lord Randolph Churchill was at odds with the dove-like Lord Salisbury, can be viewed in A. B. Cooke and J. R. Vincent, *The Governing Passion: Cabinet Government and Party Politics in Britain 1885–6* (1974), and in R. F. Foster, *Lord Randolph Churchill: A Political Life* (1981), who attributes the war to Lord Randolph alone.

11 C. L. Keeton, *King Thebaw and the Ecological Rape of Burma: the political and commercial struggle between British India and French Indo-China in Burma 1878–1886* (1974). For emigration see Command Paper C 4614 (1886), p. 165.

12 J. S. Furnivall, *Colonial Policy and Practice: A Comparative Study of Burma and Netherlands India* (1948) cf. E. E. Hagen, *On the Theory of Social Change* (1962), pp. 441–2n: "The British deposed Thibaw . . . to prevent Thibaw's negotiations with the French from developing into a situation that might cause friction leading to war in Europe."

13 J. Holland Rose, *The Development of the European Nations 1870–1914* (5th ed. 1915), pp. 529–30; cf. p. 432: "The urgent need of checkmating French intrigues in Burma led to the annexation of that land."

14 In addition to the sources quoted above see A. C. Pointon, *The Bombay Burmah Trading Corporation Limited* 1863–1963 (1964), Ch. 4.

15 IOL P 2729 India Minerals Proceedings, 1886: Finlay Fleming & Co. to Political Officer Minhla, 4 January 1886.

16 Laird, *Paddy Henderson*, p. 133.

17 Sir H. Thirkell White, *A Civil Servant in Burma* (1913), p. 303: "The development of the oilfields is the most striking feature of the economic history of the province for the past twenty-five years."

CHAPTER III

The Formative Years
1886–96

At the age of 62, David Cargill was, in 1886, to find himself more continuously busy than at any other time of his life. To place his Burmese oil operations on a sound footing, he knew that four separate things needed to be done. First, after the financial losses over more than a decade, he had to establish a limited company to attract outside funds. Second, he had to secure adequate concessions in Upper Burma from the new government. Third, he planned to bring in mechanical drilling to open up wells at a far greater depth than the 250 feet that was the limit for hand-mining, and thus raise more oil more cheaply. Fourth, he had to modernize the inefficient refining methods in Rangoon.

Limited Company

The preliminaries to forming a company did not take long once the province of Burma was incorporated into the British empire at the beginning of 1886. The Burmah Oil Company Limited was registered in Edinburgh on 7 July 1886, the authorized capital being £120,000. Cargill did not demand any promotion money and charged to the company only the barest expenses of formation. His prospectus, unusually restrained for such documents at that time, gave an account of recent oil events in Burma and judiciously forecast expected results over the next few years. Taking a "safe and moderate view of everything", he anticipated a profit of just over £10,000, equivalent to 21p a barrel as shown in Table 2.[1]

TABLE 2
Estimated profit, 1886

Receipts from products	Percentage	000 barrels	Price per barrel	total sum
Kerosene	46.0	22.1	£1.06	£23,489
Jute batching oil	26.8	12.8	0.80	10,244
Lubricating oil	1.2	0.6	2.84	1,702
Paraffin wax	6.0	2.9	3.00	8,682
Residue (used as refinery fuel)	20.0	9.6	—	—
	100.0	48.0		£44,117

Total Expenditure			%
Cost of crude oil		£14,468	32.8
Working costs (48,000 barrels @ 30.1p including transport costs of 10p)		19,575	44.4
Anticipated profit		10,074	22.8
		£44,117	100.0

Source: "Memorandum" issued with the Burmah Oil Co. Ltd prospectus in 1886

Only four directors were originally appointed – Cargill and Fleming, together with a shipowner and an iron merchant, both from Glasgow, namely Leonard Gow and James Reid Stewart. Kirkman Finlay acted as company secretary. The auditor was R. A. Murray, of the Glasgow accountants Brown Fleming & Murray, one of whose founders was a brother of M. T. Fleming.[2]

The directors' efforts to market the shares were hampered by Cargill's earlier lack of success in Burma, well known as it was to Glasgow's hard-headed business community. When in 1886

he approached some of his associates requesting them to
subscribe to the new company, many refused outright. One
dismissed him with the remark, "You won't get a penny out of
me, Cargill. I'm not going to throw away my money on another
of your hare-brained schemes." At first, therefore, he kept
going with loans, the earliest coming from his co-directors
Fleming and Stewart. He also had to revise the terms of his
original agreement with the company, taking £35,000 instead of
£40,000 for the land, buildings and plant. Only £5,000, instead
of the £15,000 earlier arranged, was paid over in cash.

Just as human childbirth is seldom without its pains, so
companies often have to struggle for life before they are safely
established. Commemorative histories of successful enterprises
usually overlook such birthpangs, which in any case soon fade
from the memory. In Burmah Oil's case, for three years after its
formation, Cargill had to continue running his oil activities
much as he had done since 1874, with only the oil supply
problem fully solved. Apart from an inaugural board meeting,
none was held until April 1889. Only then were the first shares
issued to the public, and by June just over £24,000 of the
£120,000 worth of shares had been disposed of, representing
only a fraction of the £60,000 the directors had planned to sell
initially.

Not unexpectedly after the chilly response from outsiders,
most of the earliest shareholders had direct or trading connections with the company. The directors held shares to the value
of £6,000 between them, and £2,500 was held by two prominent
shipping magnates, directors of the Irrawaddy Flotilla Company which transported the crude oil. In Rangoon, Alexander
Pennycuick of Finlay Fleming & Co. and Moolla Ismail owned
some shares.

In structure the company was merely one branch of an East
Indian mercantile house, or rather a group of more or less
interrelated businesses, and Cargill must have needed much
agility of mind to keep separate all the various strands, from tea
plantations in Ceylon to oil wells in Burma. He clearly relied
very much on Kirkman Finlay, the only one of his aides in
Britain to work full-time on oil. It must have been Finlay who
raised the question of setting up a London office.

The Rangoon Oil Company, and Cargill as its successor, had
used the address of Frith Sands & Co. at 50 Old Broad Street,

one-time importers of Burmese oil into Britain, who possessed extensive trading interests in the east. From that address Cargill had written to *The Times* to report Thibaw's succession. In 1886 Fleming moved to London to open an office in Bishopsgate, but he found the day-to-day grind not to his liking. A year or two later Finlay, having satisfactorily set up the company's secretarial arrangements, therefore came south and took this office over. He became a director in 1891, but did not travel north to board meetings unless his presence was specifically required. In Glasgow, the new Secretary was a 30-year-old employee, James Hamilton.

While the accounts, stores and plant, as well as the secretary's and share registration business remained in Glasgow, Finlay dealt with marketing, staff and technical aspects from the London office. Formally, Glasgow remained the policy-making centre until the 1920s, but from the late 1880s onwards the conduct of company affairs in Britain swung uneasily round the Glasgow–London axis, the relative influence of each end varying with the personalities involved. London was also responsible for liaison with government departments in Whitehall. Within a few years, too, Finlay was to find himself parleying with representatives of rival companies such as Standard Oil, Shell and Royal Dutch. In practice, although not in name until 1900, Finlay was the company's managing director.

Concessions

On 4 January 1886, Cargill instructed Finlay Fleming & Co. to apply again for concessions in the two fields outside the Twinzas' reservations.[3] The government of Burma duly agreed to grant a prospecting licence for 4 square miles at Yenangyaung, out of which the company would eventually be allowed to select 1 square mile for leasing. An official from the 30-year-old government body, the Geological Survey of India, demarcated the square-mile blocks on which the permits were to be issued.

The licensing system followed the exact precedent of earlier concessions in Arakan and elsewhere, granting all suitable applicants blocks to prospect and, if successful, to lease for drilling. The old promises Sir Ashley Eden had given of allowing prospectors rights over an entire field no longer held good. Licensees now had to live with the authorities' absolute resolve to avoid monopoly by encouraging the widest degree of competition between operators.

As the change of regime had terminated all former arrangements, the new government allowed Finlay Fleming & Co. to make its own bargain with the Twinzas. That was arranged between Adams, the works (that is, refinery) manager at Rangoon, and Moolla Ismail, who had bobbed up again because he was too useful to be discarded. After some protracted haggling, they reached an agreement whereby the Twinzas received Rs2½ (17p) a barrel instead of the Rs1½ (10p) paid ever since the 1850s. Moolla Ismail retained his rights to sell oil in Upper Burma, and was allowed to buy his requirements at Rs4 (27p) a barrel. All the rest would be refined at Dunneedaw and the products marketed by Finlay Fleming & Co.

British officials in Rangoon, preoccupied with reorganizing and pacifying the still very disorderly Upper Burma, duly accepted the agreement. They had already fixed the royalty at 8 annas (or about 3p) a barrel on all oil leaving the fields, the same royalty as the Twinzas had paid before 1886. The main concern of officials was to keep up royalty payments. They reported to the government of India that for the moment there seemed little prospect of getting any larger revenue without running the risk of killing the industry: a problem that has perplexed quite a few host governments since then.

Not until a new and less amenable chief commissioner had taken office in 1890 did anyone sit down to calculate the financial implications of the 1886 agreement. Those suggested that the Rangoon government was being done out of Rs220,000 (nearly £15,000) a year, some two-thirds going into the capacious pocket of Moolla Ismail.[4]

The disclosure of the large sums being made, apparently at the government's expense, undoubtedly helped to cloud the company's relationships with officials for some years to come. Its directors would no doubt have argued that for the time being Moolla Ismail was the only person who could make sure that the Twinzas maintained the flow of oil deliveries that the refinery required. In any case, his role would become largely superfluous once oil began to appear in quantity from mechanically drilled wells.

What that agreement did was to allow the refinery to function at or near full capacity. Yet for another two years much of Upper Burma remained in a disturbed state. The oilfields were from time to time overrun by dacoits, or bandits, several of the company's employees being killed or wounded, so that it had to mount armed guards over the wells. Those disturbances among other things held up both the prospecting work and the initial drilling at Yenangyaung. Even so, production in 1886 rose from the 1885 level of 45,000 to 50,000 barrels and in 1887 to 57,000 barrels.

Coolies hand-and-foot pumping a well.

Drilling

It was in the novel activity of oil drilling that Finlay Fleming & Co. encountered their most difficult initial problems, since they had no technical expertise to call on. They applied to the government of India for a man known as a "deep borer", only to find on his arrival that his skill was in boring for coal. That may have been the culminating blunder of Adams, the works manager, precipitating his dismissal in 1889. Kirkman Finlay long remembered the penalties of experimenting with incompetent men such as Adams and throwing money away in consequence.

The first driller to be engaged was a Canadian, for modern drilling techniques had been developed in North America. The earliest wells in Canada had been opened in 1861 only a year or two later than in the United States, but each of the two countries had evolved separate methods of drilling. The Canadians used a percussion system with rods and the driller struck oil in March 1889 at 727 feet. Before 1895, however, the company changed over to the American system of rope cables, which was quicker and more flexible and dealt more effectively with an emerging problem, the caving in of wells.

By the end of 1891, thirty-one producing wells had been established, and from then onwards the number of new wells opened up each year reached double figures. At Yenangyat, where the company began operations in 1891, as soon as the government had completed its survey and granted concessions, drilling was carried much deeper and the first oil was discovered early in 1893. By 1896, the company was obtaining about a quarter of its drilled production from Yenangyat.

For technical help, reliance was placed on an American named Daniel Dull who was an oil consultant in New York. How the company became associated with him is not clear, but in 1889 he contracted to supply it with drillers and purchase the requisite tools and equipment. As the scale of oil activities in Burma expanded, the directors constantly pressed him to look out for skilled men capable of undertaking deep drilling and well-deepening.

The general fields manager was William Seiple, an Ohio man who went east in 1890 and was afterwards joined by his brother Thomas and other members of his family. Finlay Fleming & Co. reckoned to interfere very little in day-to-day affairs as long as the drillers maintained a satisfactory output of oil. Even so, Burmah Oil was anxious to keep an overall supervision on what was happening. In 1894, Dull and Seiple were summoned from opposite sides of the globe for a conference with Finlay in Britain, and later with the board as a whole, about drilling problems and prospects.

Refining

Once concessions had been granted and what looked like satisfactory drilling methods had been devised, Cargill was able to give thought to his next requirement: overhauling the refinery. The Dunneedaw works had a mixed bag of equipment, some items being more than twenty years old. The twenty stills were of various types but all in separate units heated by individual brick furnaces. The refinery's condition at that time was later described as "more or less chaotic". Basic techniques were not all that much better than those used in Price's three decades earlier, with each still being charged to its full capacity no less than three times, about half being distilled away during each operation. The three distillates were mixed together to make kerosene which, as forecast in the memorandum of 1886, accounted for about half the yield. The remaining heavy oil, after cooling and pressing, yielded yellow paraffin wax and the "blue oil" used for batching jute and for lubricants. The heavy and tarry residue, known by the Russian name of *astatki*, was blown out over sawdust and rice husks, to be burnt as fuel in the works.

Cargill and Finlay insisted on the works being completely rebuilt before new plant was installed. In 1890 the newly appointed works manager, Henry Sutherland, helped by his assistant, Andrew Campbell,[5] dismantled and reassembled it on more systematic lines without interrupting output. This required great skill and energy, especially as no fewer than three creeks ran into the site, which was 6 feet below the road level.

The composition of the new plant reflected changing trends in the industry. The original techniques, which derived from de la Rue's and Price's era, were giving way to those evolved in Scotland for shale-oil. Indeed, over the coming decades, most of the improvements the company introduced were based on the shale-oil technology in which the senior works employees had been trained. The so-called continuous distillation process patented by Norman Henderson of the Broxburn Oil Company Ltd in the Lothians near Edinburgh, was the first to involve linking up stills to achieve fuel economy. There were groups of three, the primary still taking off the 1 or 2 per cent of light naphtha which was later added to the kerosene removed in the second and third stills. That kerosene, named the Victoria brand, was still smoky and discoloured, but sold well because it was cheap.

The sweating process for wax production had been improved by Norman Henderson, who had patented his own cooling process using ammonia on the absorption principle. However, refrigeration was bound to be a difficult problem in the Rangoon climate. During the first hot season when the new processes were used early in the 1890s, they were quite ineffective in keeping down temperatures. Then the refinery turned over to carbon dioxide refrigerators, encouraged by an enterprising manufacturer who installed them on the principle of "no work, no pay". The machines did work, and the refinery was thereafter able to maintain output even in the most extreme heat.

Sutherland and, after his death in 1893, his successor Andrew Campbell, concentrated on producing as much kerosene as possible since demand seemed unlimited. To achieve that aim, the directors looked carefully at any processes that might extract still more of this particular product. In 1889 Daniel Dull informed them of a Swiss invention, involving distillation under pressure, and they did what they were to do many times in the future when faced with a major decision: they instructed Finlay to "fix the issues" by going over to the United States. Finlay was initially scornful about what he called "this wonderful invention, which if it be what Dull says, will altogether alter the position and prospects of the company. . . . In this," he continued jauntily in a letter to Glasgow, "as in all matters I am at your service and will do my best if I go to show Mr. Dull that the Old Country is not altogether degenerated, although not so cute as he is."

A slightly improbable although persistent story was that he was in such a hurry that he borrowed M. T. Fleming's yacht *Norian* which, having arrived off New York, sank in full view of sightseers. That is all too reminiscent of various fictional adventure stories of the day, although Finlay was quite capable of doing such a thing. Yet whatever his mode of transport, he was won over at once to the new process. He signed an agreement and the patent stills were shipped to Dunneedaw in 1890. Within a year, they had been exposed as a complete failure, because they were incapable of maintaining the required pressure, and the stillman in charge was back in the United States.

Instead, the company was fortunate to acquire a refining patent from Dr Boverton Redwood and (Sir) James Dewar, which allowed more of the lighter fractions to be extracted and hence markedly increased kerosene yields from 45 per cent in 1889 to 57 per cent in 1897. Redwood, then in his mid-forties, was the leading oil consultant in Britain. He was unforgettably described by a consultant of the next generation, A. Beeby-Thompson, as "immaculately attired, with an orchid in his buttonhole, speaking in a slightly affected but impressive manner with a provincial accent".[6] His imposing stature and profile often led him to be mistaken for the actor Sir Henry Irving, something that he was vain enough to encourage.

His consulting role, involving professional advice to public bodies and private companies, often took him to oilfields and refineries in America, Russia and elsewhere – although never as far afield as Burma. He was, therefore, a walking encyclopaedia of oil knowledge, some of which he put into his authoritative works on petroleum, while some he kept in his head. He chose his friends carefully; Kirkman Finlay was one of them, and it was he who persuaded Redwood to become part-time consultant to the company, a post he held from 1893 until his death in 1919.[7]

In 1888 about 65,000 barrels of crude oil had been produced in Burma, virtually all by the Twinzas. That compared with 50,000 barrels in 1886. Three years later the total had more than doubled to 144,000 barrels, over half coming from the company's own wells, so quickly had the Twinzas lost their predominant role in oil output. By 1892, the refinery could no longer cope with this flow of crude oil, and a further extension was essential; but the works manager unimaginatively suggested duplicating Henderson's stills one by one as required.

Finlay would have none of that. He argued for a really large extension which would double capacity and achieve the maximum economies of scale. Enough crude was already available, or at least within call, he maintained, to meet every demand. When Pennycuick from Rangoon asked how they could possibly dispose of the extra production from the enlarged refinery, Finlay stated roundly – somewhat like the Scottish preacher encountering religious doubts – that problems of that kind would "disappear on being confronted."

The Glasgow board was clearly flummoxed by the whole debate. It discussed the issues inconclusively a number of times that year, and then ordered Finlay out to Rangoon to assess the refinery's needs on the spot, taking account of estimated future crude supplies. Setting off without delay, he arranged also to meet the chief commissioner and officials in Rangoon to plead for more concessions. On his return he persuaded the board to authorize an interim extension that would allow output to be kept up until a thorough overhaul could be carried out. But he continued to worry about the refinery's future, and above all what might happen should it burn down, since the company now had sales contracts carrying severe penalties for non-delivery, particularly of jute batching oil in Calcutta, the next best profit-earning product after kerosene.

Early in 1895, therefore, he steered through the board plans for building an additional refinery on an entirely separate site, chosen by himself during his visit. Capacity would at first be half as much again as at Dunneedaw. It was on the opposite bank of the Pegu river, at Syriam; thus in the event of a fire or explosion it would be far less of a hazard than the old refinery, which was close to Rangoon city.

Finlay may have been sublimely self-confident about being able to market the various refinery products, but matters were not quite so easy as he supposed. Of the kerosene, between a half and a third currently found a market within Burma itself and the rest was shipped to India, mostly to Calcutta. As already shown, the trade was a long-standing one. Ever since 1871, oil from Rangoon had been sold in India through the Calcutta mercantile house of Wiseman Mitchell Reid & Co., but only relatively small amounts were involved. The 11,000 barrels a year sent to India by the Rangoon Oil Company in the early 1870s had risen only to 12–16,000 barrels a year in the early 1890s. By then 700,000 barrels each were coming from the United States and Russia. The agent therefore earned negligible profits from sales of Burmese oil and found its marketing an uphill struggle even though in 1894 the recently imposed tariff on imported oil was doubled, to represent a 14 per cent advantage to the company.

The refinery staff at Dunneedaw would dearly have liked to make a superior brand to match the colourless and clear American product. Unfortunately, the state of technology did not permit the waxy crude coming from Yenangyaung to be made to yield a satisfactory product. Again and again the company sought refining standards that were technically out of its reach. Not until 1905 could it put on the market the first of its superior brands, Gold Mohur.

Thus Burmah Oil's kerosene trade with India continued to be of minor consequence for some years to come, but from 1891 onwards the agency was transformed when it was taken over by R. G. Shaw & Co. of London, the Calcutta branch being Shaw Wallace & Co.[8] That brought Finlay into contact with two of the new agency's principals, who were destined later to play a key role in the development of the company and its even more considerable offshoot, the Anglo–Persian Oil Company. One was the senior partner, Charles W. Wallace, then aged 35 and resident in London, and the other Charles Greenway, a year or two younger and based in Calcutta since 1893. Remembered as "singularly able, industrious and resourceful young men", they and their activities and opinions figure largely in the pages that follow.[9] Finlay had been concerned enough about the Indian market to pay a rushed visit to Calcutta during his trip in 1892, meeting there the managing partner of Shaw Wallace & Co., H. S. Ashton, and instilling in him some of his own infectious enthusiasm for the prospects of oil.

It was its next most important product, jute batching oil, which was used to soften the fibres in preparing jute for processing, that first brought the company into contact with the Standard Oil Company, the giant American corporation founded by John D. Rockefeller, and by then the most extensive business enterprise in the United States.[10] Before 1886 Burmese jute batching oil had taken second place in India to the Scottish variety, but by 1895 had caught up. Then Standard Oil approached Finlay with an offer to divide the market and keep a third for itself, allowing two-thirds to the company, which would presumably have to share the market with its Scottish rivals as best it could.

On hearing the offer, the directors took it seriously enough to summon their general manager, Pennycuick, from Rangoon to London for consultations with Finlay and with Greenway, then

on leave from Calcutta. They finally agreed to accept Standard
Oil's terms; at that stage nothing could be gained by antagoniz-
ing such a powerful rival. Lubricating oil gave no problem, as it
could be sold to the Irrawaddy Flotilla Company for use in its
vessels, and to the Burma Railways. Paraffin wax, the only
other marketable product, is discussed in Chapter IV.

The directors' satisfaction over progress to date was rather
soured by the activities of certain rivals who had sprung up in
Burma. The first person of any consequence to apply for a
concession was Sir Lepel Griffin in 1889, as head of an oil
syndicate.[11] He had done well in the Indian Civil Service,
although his flamboyant but unstable character had kept him
out of the top posts. Lord Dufferin invited him in 1886 to take
the unrewarding job of reorganizing Burma, but he refused,
and when he was passed over for an Indian governorship, he
retired in 1889. With typical exuberance, he initially sought
rights over the whole oil-bearing territory in Burma, but then
settled for 4 square miles, which he was granted early in 1890.
    When the *Rangoon Times* broke the news under the headline
"Liberal Concessions" Cargill, through Finlay Fleming & Co.,
protested to the government of Burma that his own company
was being given second-class treatment, and that they had
totally failed to recognize his expenditure of no less than
£100,000 and all his efforts over the previous thirteen years.
Moreover, some of the blocks already promised to him were
being allocated to others. The protest went home, and he
received an additional total of 2 square miles together with a
promise of "the pick of the field" in advance of Griffin's
syndicate. That July he protested to the India Office in London
that since 1874 he had relied implicitly on "distinct promises
and assurances" given by successive chief commissioners,
pledging full support and encouragement. Thus his current
treatment seemed to afford "just grounds for strong remonstrance".
    His complaint set off a first-class row in Whitehall. A spate of
official minutes, telegrams and letters all condemned the government of India for the embarrassingly generous terms offered
to Griffin at the expense of Cargill's strong claims as pioneer of
the oil industry in Burma, and contrary to the advice of the chief
commissioner of Burma. Doubtless Griffin, with his eminence,

strong personality and intimacy with the government machine, had prevailed on some middle-ranking official in Calcutta to make the unfortunate decision. Higher authority there did everything possible to minimize the damage; for instance, someone floated the idea that the Secretary of State for India could use his powers to veto the award, a suggestion indignantly rebutted in Whitehall as a ruse to throw the odium onto the Secretary of State rather than where it lay.

Meanwhile Cargill was energetically pursuing the justice he demanded. Having spurned the consolation prize of 4 square miles in Minbu – where oil prospects at that time were not at all encouraging – he and Finlay requested a high-level meeting at the India Office. There they refused to be fobbed off, and hinted that in the last resort they might have the matter raised in the House of Commons. That hint worried officials, already embarrassed by parliamentary questions about military excesses

Thatched derrick, Yenangyaung oilfield.

over the pacification of Burma. The Under-Secretary of State, Sir John Gorst, privately favoured an official investigation, being convinced that there had been a "job" (meaning corruption), but his minister ruled that out, instead blaming imprudence. In the end Burmah Oil was granted parity with the syndicate, namely 4 square miles at Yenangyaung. The directors need not have worried. Test drillings by the syndicate produced nothing and it soon ceased its operations.

Another successful applicant was Charles Aria, inventor of a type of lamp based upon a German model, for burning poor-quality kerosene. He spent much money in trying to prove his concession, and planned to combine with other licensees to form a company to be named – rather cheekily – the Burma Petroleum Company. His potential associates included Alfred Suart, mainly active in the Russian fields, and one of the many colourful characters who adorned the oil scene in its pioneering days.

Finlay for one was convinced that Aria and his associates were not really serious in wanting to discover and refine oil, but were merely speculators. When some of them had called on him in 1891, putting out feelers for a tempting offer to buy, he had been distinctly unforthcoming: any proposals they might wish to make, said Finlay, would receive most careful consideration. Not until 1896 was a company, the Minbu Oil Company, formed. Its fortunes will be related in Chapter IV.

By 1895, active opposition to Burmah Oil seemed to have been contained for the time being. Considering the progress it had made in various directions during its first decade, the chairman was justified in using the word "satisfactory" a number of times in his report for that year. It had been a good one for Burmah Oil, showing a "steady and gratifying increase in the extent of the work done and the results obtained", he thankfully recorded.

Notes

1 The "Strictly Private and Confidential" prospectus of 7 July 1886 was accompanied by a memorandum, which is the source of the figures quoted in Table 2.
2 For R. A. Murray see E. Jones, *Accounting and the British Economy 1840–1980* (1981), especially pp. 85–9.

3 IOL P 2729 India Minerals Proceedings, 1886: Finlay Fleming & Co. to Political Officer Minhla, 4 January 1886.

4 IOL India Minerals Proceedings, 1891: H. Thirkell White to Secretary of Government of India, 9 February 1891.

5 For obituary of A. Campbell (1868–1941) see *Journal of Institute of Petroleum*, 27, 1941, pp. 310–11.

6 A. Beeby-Thompson, *Oil Pioneer* (1961), p. 80.

7 An important article by Redwood, 'The Oil-fields of India', is in *The Journal of Society of Chemical Industry*, IX, 1890, pp. 359–70, for which Finlay provided data.

8 The early history of the Indian agency is by H. S. Ashton in Sir Henry Townend, *A History of Shaw Wallace & Co. and Shaw Wallace & Co. Ltd.* (1965), Ch. I.

9 Ibid., p. 1.

10 In 1870 the 31-year-old John D. Rockefeller incorporated the Standard Oil Company in Cleveland, Ohio, where he had acquired control of a refinery in 1865. Standard Oil companies were soon set up in a number of US states. In 1882 the Standard Oil Trust was formed so that control of the companies could transcend state boundaries. The Standard Oil Company of New Jersey (see Table 1) dominated the group after 1888 as New Jersey's laws permitted cross-holdings of stocks between companies in different states. A historic anti-trust lawsuit in 1911 forced the division of Standard Oil into more than thirty separate independent companies. Standard Oil of New Jersey remained the most important of these entities. Its British marketing affiliate was the Anglo–American Oil Company Ltd, founded in 1888. Standard Oil's wholly-owned subsidiary, the Colonial Oil Company, unsuccessfully sought licences in Burma in 1902 (see ch. V).

11 The Griffin episode can be followed in IOL L/E/7/223-5, 228 and 230, 880/90, 932/90, 1102/90, 1326/90 and 1473/90 respectively. The final reprimand from the India Office to the Government of India, dated 29 January 1891, and its successive drafts are in 1473/90.

CHAPTER IV

Turn of the Century
1896–1901

A turning point in the Burmah Oil Company's formative years occurred in 1896–7. Before then the company had been struggling for existence, with the risk of collapsing as its predecessors had done. Thereafter, short of a calamity, its prospects for long-term success were largely assured. A new atmosphere of professionalism and self-confidence pervaded the company's activities. Earlier doubts over supplies of crude were resolved, at least for the moment, when in 1896 the experts of the Geological Survey of India confirmed the richness of the Yenangyaung deposits, where successive flowing wells were drilled at depths of more than 700 feet.

In the following year an official of the Geological Survey, G. E. Grimes, mapped out the area that was to become the Singu field (later known as Chauk), also on the banks of the Irrawaddy some 30 miles to the north, thus drawing attention to its potential.[1] The company at once applied for as many leases as it could obtain in that field, and first struck oil there in 1901. By the mid-1920s, when Yenangyaung was past its prime, Singu was acknowledged to be the premier field in Burma. Meanwhile, as if to drive home the lesson that oil is a risky business, operations at Minbu in the south had from 1895 onwards become unproductive. Having carried them down to the record depth, for Burma, of 1,700 feet, the company abandoned drilling there and did not resume it until 1914.

The Yenangyat field, on the opposite bank to Singu, continued to produce an oil of much lighter consistency than at Yenangyaung, and the extensive new works at Syriam, which

46

Arakanese well, Ramree Island, typical of the early primitive wells.

came on stream in 1897, were therefore designed to refine that lighter oil. However, deliveries from Yenangyat were well below the refinery's capacity and the shortfall was made up with Yenangyaung oil. In the event, Yenangyat reached its peak as early as 1903 and yields thereafter began to decline. Even so, Syriam was established as the growth-point for refining, and it was extended no less than ten times between 1898 and 1909.

Whereas in 1889 only 60,000 barrels had been refined, by 1895 – before Syriam was completed – the throughput had risen to over 160,000 barrels and by 1901 to nearly 760,000 barrels, as Table 3 shows. Finlay Fleming & Co. therefore kept up constant pressure on the government of Burma to extend the area over which the company could take up drilling leases and by 1898 it was allowed a maximum of 20 square miles.

TABLE 3

Crude production and refinery throughput 1889–1901 (000 barrels)

	Production				
	Yenangyaung		Yenangyat		
	Hand-dug	Drilled	Hand-dug	Drilled	Total
1889	67	4	1	—	72
1895	122	182	1	23	328
1901	156	765	1	293	1,215

	Refinery throughput and yield					
	1889 (a)		1895 (a)		1901 (b)	
Throughput	60		163		758	
Products		(%)		(%)		(%)
Naphtha	—	(—)	—	(—)	81	(11)
Kerosene	27	(45)	84	(51)	429	(57)
Jute batching oil	17	(28)	40	(25)	55	(7)
Lubricating oil	1	(2)	5	(3)	9	(1)
Paraffin wax	3	(5)	12	(8)	23	(3)
Fuel oil	—	(—)	—	(—)	84	(11)
Residue	12	(20)	22	(13)	77	(10)
Total	60	(100)	163	(100)	758	(100)

Notes: (a) Dunneedaw only; (b) Dunneedaw and Syriam
(Burmah Oil Records)

The increase in yields came almost entirely from drilling. In the mid-1890s only about half the crude oil raised was actually turned into refined products. About an eighth was consumed by the Twinzas, and perhaps another quarter was required in the oilfields to power the many steam engines that worked the drills and later pumped the wells. Between a quarter and a sixth was used up as fuel in the refineries; this was a high proportion which, although shrugged off at the time, later caused the directors much anxiety as they began to realize that the supply of crude was not limitless.

Kerosene continued to be the most important of the company's products at a time when technical improvements, notably the Redwood-Dewar process, were satisfactorily stepping up yields of the Victoria brand. Although its quality could not be improved in the prevailing state of technology, it was at least readily saleable and the surplus for export to India mounted year by year. In 1896/7 it reached 20,000 barrels, which was still only 1 per cent of India's total consumption; by 1901/2 it had risen to 337,000 barrels, or about 13 per cent.

In common with other expanding companies, Burmah Oil found that such rapid growth put pressure on ancillary services, notably sea transportation. Maritime regulations limited the amount of oil products any one vessel could carry, and then only as deck cargo: the containers for kerosene were 4-gallon tins, two comprising a "case" or wooden crate. Transport was therefore expensive, added to which, as Finlay reported candidly to the board in September 1898, the British India Steam Navigation Company was incapable of providing a reasonable service when its steamers were in a constant state of pressure and hurry from port to port. The tins themselves were so roughly handled that leakage often occurred with consequent financial loss and the forfeiture of goodwill.

His remedy was for Burmah Oil to acquire its own tankers.[2] The company's bill for sea transport was then about £12,000 a year and all set to soar as exports increased in line with refinery extensions. Two tankers costing £30,000 each, plus all the operating costs, would save a total of £5,000 a year and bring to an end leakages and damage to cargo.

The board duly agreed to order two 2,000-ton tankers, capable of shipping refined products in bulk or as cargo in tins and drums. The ships ran on fuel oil and the pumping system was of novel design. Unfortunately, the still worthless naphtha, after extraction, could no longer be mixed with the kerosene and had to be put into the fuel oil instead. That reduced the flash-point to a dangerously low level, and a disaster was merely a matter of time. During its maiden voyage, one of the new tankers, the *Kokine*, caught fire at Rangoon and became a total loss. The directors at once ordered a 2,500-ton replacement, the *Khodaung*. Until 1905, when at last it could be used productively, naphtha was burnt off.

Finlay had to cope as best he could with quite novel problems. Suitably qualified masters for tankers were almost impossible to find, and he turned to the elder James Galbraith's firm of P. Henderson & Co. for advice. The upshot was that Captain George Currie was given command of the other tanker, the *Syriam*. A Paisley man, he was an old salt brought up on sail, full of self-confidence over practical matters but utterly at sea when he had to contend with paperwork.

Not surprisingly, Finlay gave him precise instructions on how to handle his command. On the outward journey, "take no risk at all", Finlay ordered. "The *Syriam* must arrive safe and sound. If necessary heave to in the Med. for a time." Later he was to reprimand Currie for having run the tanker too fast for safety, bearing in mind the volatile nature of his cargo, and for having gained his experience of such vessels at the company's expense, presumably through making various costly mistakes. However, they soon recognized his high quality as a skipper, and in due course appointed him as the company's first marine superintendent.

The company's troubles were by no means over when the products reached the port of discharge. With his usual foresight, Finlay had already acquired some land on lease at Budge Budge in the Calcutta docks area; there the 500,000-gallon storage tanks and a factory to turn out 5,000 4-gallon tins –

Burmah's first tanker *S.S. Syriam* (2088 tons dwt) launched at Grangemouth, Scotland in 1899. Note awnings rigged against the heat of the day.

equivalent to 500 barrels a day – were erected, in time for the *Syriam* to deliver its first bulk cargo there in January 1900. The following year Shaw Wallace & Co. opened an agency for Burmah Oil at Bombay; similar agencies followed at Madras and Karachi during 1906 and 1907 respectively and at minor ports later on. The company therefore added three more tankers to its fleet, the *Twingone*, *Singu* and *Beme*, between 1902 and 1904. These had more than double the displacement of the earlier ships, namely 4,700 tons deadweight. Between them the five vessels carried practically the whole of the company's ocean trade until 1914, helped by two smaller vessels for the coastal trade.

Finlay's efforts to provide the company with its own tankers and tank installations were not merely to make financial savings, but also to reduce dependence on others, at a time when the size of the potential Indian market seemed limitless. He always kept in mind the company's strategic interests. Soon, he believed, rival firms – especially under the leadership of Standard Oil – could well conspire together to force up oil prices and divide the market according to fixed shares. By creating efficient and self-contained transport and storage facilities, Burmah Oil would be strengthening its hand in any future contest with the oil giants. Meanwhile, the company's aim must be to avoid being trapped by all or any of its powerful rivals.

For advice on his overall strategy he was regularly in touch with the company's consultant, Boverton Redwood. Through his eyes we can see how, early in 1897, the foremost expert in Britain viewed the world oil scene. Redwood's forecast was that oil consumption would soar every year and that drastic changes to the oil supply pattern of the world as a whole would occur within the next few years. While a few oil companies then dominated global markets by distributing refined products world-wide, Redwood expected that pattern to alter once local, or what he called "sub-divisional", sources of output sprang up.

American oil, he reckoned, would increasingly be earmarked to meet domestic demand and therefore no longer be freely available for export, particularly to the east which was its remotest and least lucrative market. Similarly, production in Russia would be gradually absorbed by that vast and industrially developing country and its immediate neighbours.

Since the United States then accounted for half the world output and Russia for 45 per cent, Redwood assumed that all the prospecting in progress would discover new fields in widely scattered parts of the globe to redress the production imbalance. As he stressed, everything depended on competition being maintained in production and refining. His views in 1897 were not entirely disinterested, for he was then seeking to persuade Finlay to buy up some oil interests in the Indian province of Assam. Finlay's response reveals how far he already dominated company policy. David Cargill as chairman was enthusiastic about going into Assam, only to be overridden by Finlay who believed in concentrating rather than dispersing the company's efforts.[3] As he explained to Redwood, "we had already quite as much in our hands as would occupy us for the rest of our lifetime."

Redwood's prophecy that oil production would in due course be more equally distributed throughout the world, was not to be fulfilled during the period covered by the present book: by the mid-1920s, and indeed until the Second World War, over four-fifths of global production came from the western hemisphere. This was never to be known to Finlay, whose worries about the power of Standard Oil were more immediate concerns. For years that company had looked upon India as its largest single overseas market outside Europe, and was likely to resent large-scale incursions into the Indian market.

Finlay knew precisely what he was up against with the Americans. "Anything," he confided to Glasgow with stark realism, "would be better than having the Standard Oil Company arrayed against us. Nobody has done that for many years that is now living – commercially speaking." Yet when in 1897 an emissary from Standard Oil proposed to him a joint price war "to knock the Scotch [shale-oil] trade on the head", he stood up to him and won the right to have a reasonably high prior share of the jute batching oil contracts with the Calcutta jute mills. Having successfully claimed the market as his own, Finlay managed to hold on to it until 1901, when he had to reach a more restrictive agreement with Standard Oil, which barred him from turning as much of his expanding refinery throughput into jute batching oil as he would have liked. He had to hold back his paraffin wax production as well, and join with Standard Oil – the world market leader here – and the

Scottish firms in accepting a price structure. A decade or so was to pass before a comprehensive wax pool agreement for India was reached (see chapter XI).

At the end of the 1890s, Burmah Oil made its first contacts with the two producers in the Dutch East Indies closest to its own fields in Burma. The Royal Dutch Company, founded in 1890 to develop a concession in Sumatra, had made rich oil strikes there and by the middle of the decade had achieved a kerosene production that was some four times as great as Burmah Oil's.[4] At first that company disposed of its kerosene in far eastern markets well away from the Indian empire.

The Shell Transport and Trading Company Ltd, on the other hand, was likely to prove a major rival. As its name suggested, it was originally concerned with oil freight, its founder Sir Marcus Samuel at first shipping Russian oil through the Suez Canal to supply his eastern markets, including India.[5] In 1896, a year before Shell became a limited company, he began drilling in the island of Borneo. By then Finlay was attempting to get to know him through various intermediaries. "We are most distinctly anxious that they [the Shell people] should write us down as friends" was the burden of his messages.

Not until May 1900 did the two men finally meet, after Samuel had complained about some kerosene being shipped from Rangoon to the Straits, i.e. Malaya. Those shipments, trifling as they were, had succeeded in disrupting the market by forcing some price reductions on Shell. Samuel claimed the Straits as his own "pocket borough" and, to protect it, threatened to carry the war into Burmah Oil territory by erecting a tank installation in Rangoon.

"Coals to Newcastle", was Finlay's characteristically robust retort, and he added that in any case the Straits market was a very old established one for Burmese oil, dating back at least to the 1850s. Privately, however, while convinced that ordinarily "one must not take Sir M. Samuel's utterances too much *au sérieux*," he felt the latest threat to be a real one. As Burmah Oil had far more to lose by Shell obtaining a foothold in Burma than it gained out of a few shipments to the Straits, he agreed that the two companies should henceforward keep out of each other's territory.

Finlay also formulated what were to be the two principles of the company's strategy for many years to come. First, he insisted on a prior claim over certain product markets in India. Second, he refused to consider any moves to restrict output with the object of preserving artificially high prices in India. As he wrote to Greenway in Calcutta, "There must be . . . no hampering of our sales to suit the convenience of the Shell Company. That is the ABC of our policy." At the same time, he was ready to agree reasonable prices with competitors, so as to maintain orderly markets in everyone's long-term interest.

The potential threat from Shell really depended on how successful its Borneo refinery at Balik Papan would turn out to be, and here Finlay had gleaned some vital intelligence. Samuel had apparently believed that he had in Borneo an unlimited supply of saleable oil. As Finlay reported, "like other ignorant persons at the present moment he thought oil was oil as long as you got it out of the ground and that the usual stereotyped refinery would bring about the usual results." In the event, the oil turned out to be low-grade, yielding 15 per cent of a heavy and smoky kerosene, the remainder being valuable only as fuel oil, which was restricted to maritime use at that time. So Samuel "with that quick wit for which he was celebrated, instantly began to shout his war-cry of 'oil fuel for steamers!'" This was of little concern to Burmah Oil, which turned out only small quantities of fuel oil for its own tankers and Irrawaddy Flotilla steamers.

A far greater worry early in 1901 was caused by persistent rumours of a possible merger between Standard Oil and Shell. A giant oil trust of that kind would have been perfectly capable of wiping out his own company; however, there was, in his words, "so much lying going on that it's no use even trying to get at the truth." Redwood tried to calm his agitation by explaining that to the best of his knowledge relations between the presumed suitors "only amounted to a sort of mutual offensive-defensive understanding of playing to each other's hands", whatever that meant. Even so, Finlay poured out his apprehensions to Cargill in a nice clutch of mixed metaphors:

> Samuel wants the Burmah Oil Company to do – or not to do – something or other, and throws in his trump card played to draw our fire. Now we are on dangerous ground, and must play our cards

with great firmness and ability. Neither you nor Fleming will have forgotten my views in re the "coming storm" in oil in the east, for I've "dinned" it into all your and Rangoon's ears for years.

I wish the Burmah Oil Company were two or three years older and bigger, before being called on to fight. . . . The question now is – are we to draw Samuel or wait for Samuel to draw us? If the former, we must do it pluckily.

True to Samuel's earlier threat, Shell instructed its Calcutta agents to apply for a site at Rangoon to erect oil tanks.

Finlay, having interpreted the application as the first shot in a developing contest, initially thought he would "bide a wee". Then he changed his mind and decided to argue it out with Samuel face to face. On reaching the Shell offices he was astonished to find two separate meetings being held in the chairman's room. Once able to gain a hearing, he treated his call in a "light and airy way", speaking very deliberately and slowly as he might to a recalcitrant animal, child or foreigner, and giving Samuel plenty of time to think before answering. The response turned out to be totally unexpected; Finlay could only liken it afterwards to the collapse of the "walls of Jericho". Shell's plan to erect storage tanks was not a reprisal, Samuel explained, and he would agree not to erect them or "pin-prick the company in any way" on Burmese territory.

Although Finlay was not to know it, Samuel had far more weighty business on his mind, nothing less than a bid to secure for Shell the Texas oilfields in the United States which had become productive at the beginning of 1901. He was thus more than content with having squeezed Burmah Oil out of the Straits. He did not bother Finlay again for almost a year, and about the only flurry from that direction occurred in June when the company secretary of Shell, in a mysterious telephone call, claimed that the Rothschilds had had a whisper about the intention of Standard Oil to buy up Burmah Oil. "If true, hold your hand," that individual coolly advised Finlay. "We will go one better whatever Standard Oil may bid."

The astounded Finlay could only rejoin that even if Standard Oil had the cash – which he doubted – his own company was quite definitely not up for sale. "The present owners of the Burmah Oil Company Limited know the value of their property and are keen to keep it," he declared loftily. Then early in

December the rumours were confirmed that Standard Oil was trying to acquire Shell. As Shell would have ceased to exist as a separate entity, Samuel finally rejected the Americans' bid just before the end of the year: no doubt to Finlay's profound relief.

Meanwhile, Burmah Oil continued to consolidate itself. For 1899, profits remitted from Rangoon came to £96,500, over three times the average for 1894–5. That represented an encouraging 24p per barrel and allowed a record dividend of 25 per cent to be paid. Finlay expressed his feelings of thankfulness to Cargill, but urged him to keep the results very quiet because there was "a lot of gossip knocking about" which would only "attract flies to the sugar". In the least felicitous phrase he ever used, he advised, "let 'lockjaw' be our motto." However, sealed lips could not be guaranteed to deflect the intelligence networks of oil rivals, and it was no coincidence that from then on they began to pay increasing attention to the company.

In his turn, Finlay directed his attention to rivals within Burma itself. As he informed Glasgow:

> It must be our aim to frustrate as far as we can any additional oilfields or oilworks that may oppose us. We have in Burma, I believe – even if our capital were multiplied by four – just as much as our energies and capabilities can handle during our lifetime, and we must concentrate, not dissipate, those energies.

That was the more necessary as he anticipated "astonishing changes" shortly in the world of oil. His particular target in the province was the Minbu Oil Company, founded in 1896, with a number of blocks in Yenangyat.

The main promoters were all partners in the house of Ogilvy Gillanders & Co., Henry Neville Gladstone, son of the former prime minister, and two distant Gladstone relatives. Two years later, having discovered oil in their blocks, they proposed to expand their business, relying on hints about official reluctance to award any further concessions to Burmah Oil. The ever-tactful Redwood thought it advisable to bring Finlay and the Gladstones together in London. That initial meeting could not have gone off worse. Finlay was anything but impressed by "these gentlemen" with their "jubilant and jaunty" attitude which in his eyes verged on "arrogance and unreasonableness". He was convinced that, like many rivals before and later, they simply wished to sell out to Burmah Oil at a profit to themselves. "Let them stew in their own juice till they come to us," was his contemptuous message to Glasgow.

It was not long before they came to offer their concessions and plant for £40,000. Not a penny over £30,000, was Finlay's reply, and it took a month's hard bargaining before they finally agreed to split the difference. Finlay could not resist giving himself a small pat on the back but he assigned the major credit to Redwood, the company gladly paying the £3,000 fee due to him from the Minbu Oil Company: "the price," in Finlay's appreciative words, "due to the prince of oil diplomatists."

His gratification was rather clouded by some fussing from Glasgow about possible official reprisals once the news of this frankly monopolistic take-over leaked out. As he urged on Pennycuick, it was vital to ensure that nothing should leak out. "Guard the secret of change in the Minbu Oil Company religiously," was his conspiratorial advice; "let there be no mistake about it. I admit nothing to anyone."

That agreement had a further potential bonus. In 1897, another rival syndicate had been formed jointly by J. H. Moore, an American driller who had been dismissed by Burmah Oil, and S. A. Mower, a Scotsman long engaged in the rice trade. However, although the fields manager, William Seiple, opened negotiations at the oilfields with the syndicate's representatives, they proved extremely elusive and evaded the company's clutches. Four years later, they incorporated their syndicate as the Rangoon Oil Company, a name which Cargill had unfortunately allowed to lapse.[6] Its rupee capital was equivalent to no more than £28,000.

Despite these parleys with rival well-owners and organizations in the province, Burmah Oil's prosperity rested ultimately on the willingness, or otherwise, of officials to grant licences and their readiness in case of need to protect the company against foreign incursions. In 1898 the directors, through Finlay Fleming & Co., officially requested the government of Burma to increase the concession limit from 20 to 100 square miles. There was a need to look far ahead, and in any case, the company felt that it deserved some consideration for having spent, since 1888, about £200,000 on plant and wages in the Burmese oilfields.

Behind the scenes, as the official minutes show, the government of Burma was far from happy about the growing giant in its midst. Burmah Oil seemed to be getting a great deal too affluent; besides, an extension to 100 square miles or more would give it a "practical monopoly", which it might exploit to force a reduction in its royalty payments. Officials still saw the issue purely in financial terms. They looked on the company as a revenue raiser and disregarded the fact that crude oil was one of the province's major resources, calling for well-defined policies of orderly development and conservation. Finlay would not have been reassured to know that, in a discussion at the India Office at about that time, the Permanent Under-Secretary told his Secretary of State that he was unhappy about the Burmese oil industry being artificially protected. He grudgingly accepted the need for tariffs until such time as the oil industry in Burma was fully developed and could "stand alone": thereafter, he felt, an excise duty equivalent to the tariff should be imposed or the royalty raised accordingly.[7]

By the end of 1899 the authorities did relent and awarded an extension of the limit to 50 square miles, but the lieutenant-governor – as the chief commissioner had become – was no friend of the company and reserved certain blocks against it: that is, he denied them to the company while making them freely available to other applicants. No act could possibly have annoyed the directors more, and they complained on many subsequent occasions. Thus in Yenangyat the company's blocks were often surrounded by those of rivals, who were able to do their prospecting on the cheap by watching for the results when Burmah Oil put down its drills. If oil were struck, then the rivals began to drill as quickly as possible to draw off the maximum amount of oil from the vicinity. The authorities failed to grasp that problem early enough, and, as Chapter VIII will show, not until 1908 did they impose strict regulations about well-spacing and other matters with the object of achieving orderly development in the fields.

The company tried to justify its high profits on the grounds that the producing wells were highly variable in their yield and that the Burmese fields were difficult to work because wells were always liable to cave in and flood with water. All these basic problems about oil reserves and threats to the company from within and outside Burma weighed heavily on Kirkman

Finlay, who was only too aware that it badly needed a period of consolidation, if possible lasting some years, that would involve no drastic or costly changes in direction. How far events permitted of such a fallow period remains to be seen.

Notes

1 G. E. Grimes, *Geology of Parts of the Myingyan, Magwe and Pakokku Districts, Burma* (Memoirs of the Geological Survey of India, XXVIII, Part I, 1898), pp. 30–71.

2 For the company's tanker fleet 1899–1962, see H. H. Twist, "The First Fleet", *Burmah International*, 21, January 1975.

3 For the early days of what later became the Assam Oil Company see *The Story of the Assam Railways and Trading Company Limited 1881–1951* (1951), ch. VIII, on "Oil" and Redwood's article mentioned in ch. III, note 7.

4 Until 1907 Royal Dutch (Koninklijke Nederlandsche Maatschappij, founded in 1890 to develop the oil deposits of Sumatra) and the Shell Transport and Trading Company Ltd (founded in London in 1897 by the Samuel brothers) were independent companies, although in 1903 they merged their marketing organizations in the eastern hemisphere into the Asiatic Petroleum Company Ltd. In 1907 the Bataafsche Petroleum Maatschappij was registered in Holland to control production and refining and the Anglo-Saxon Petroleum Company Ltd in England to control the tankers and tank installations; Asiatic continued its marketing role. Royal Dutch and Shell thereupon became holding companies, the former owning 60 per cent of the operating companies and the latter 40 per cent. Although in practice relations between the Dutch end (which concentrated on technical aspects) and the British end (which dealt with financial and commercial affairs) were remarkably harmonious, the existence of Dutch voting control over these world-wide assets led in 1915, during the First World War, to planning for British control through a merger with Burmah Oil (see chs XVI and XIX). In the absence of a fully documented history of Royal Dutch-Shell (which is now badly needed), an outline of events can be found in K. Beaton, *Enterprise in Oil: A History of Shell in the United States* (1957), ch. II.

5 A good and sympathetic biography is by R. Henriques, *Marcus Samuel: First Viscount Bearstead and Founder of the "Shell" Transport and Trading Company 1853–1927* (1960). Samuel's creative qualities, particularly in earlier years, must be borne in mind as a counter-weight to the rather blustering attitude portrayed in the present work. He subsequently allowed himself to be manoeuvred by Deterding into a position of relative impotence in Shell's management.

6 David Cargill had acquired the assets of the original Rangoon Oil Company in 1876 (see Introduction and ch. II).

7 IOL L/E/7/441, 3436/00: Sir A. Godley to Lord G. Hamilton, 10 December 1900. Hamilton, on the other hand, disagreed: "We ought to foster our Indian industries by all means instead of applying to them the rigorous doctrine of free trade."

CHAPTER V

Reconstruction
1901–3

In the first years of the twentieth century, the stature of the
Burmah Oil Company continued to increase as its activities
began to impinge on world oil affairs. Hence the directors saw a
need for certain internal reforms, notably to strengthen the
company's financial basis. This was done in the "reconstruction" of 1902, and Kirkman Finlay's subsequent efforts further
to reduce its dependence on others led to more tanker building
and the first serious plans for a pipeline.

When Finlay studied the accounts for 1900, he was more
than ever concerned at the cost of transporting crude oil
down-river to the refineries. He had inside information suggesting that the Irrawaddy Flotilla Company's directors were
deliberately burying their profit in hidden reserves and "sing-
ing very small" with the object of extracting high rates out of
Burmah Oil, by far their largest customer. "They want the
plunder," he wrote melodramatically to Glasgow, "but as I live
they shall not have it."

He also deplored the Flotilla Company's practice of "working on the cheap" by refusing to increase its fleet of river
steamers. This had already endangered the flow of oil on
several occasions, all but bringing the refineries to a standstill.
He had admittedly used hard bargaining over the years to bring
down the rates per barrel as annual oil shipments had soared.
The latest contract had nearly caused a serious row in Burmah
Oil, largely because the Flotilla Company's managing director,
John Innes, had joined the board after James Reid Stewart's
death in 1896. Finlay was unhappy about having Innes as a

Irrawaddy river boat.

fellow-director because the Flotilla Company was the one organization that Burmah Oil needed to keep at arm's length. Moreover, he believed that Innes was at times indiscreet to outsiders about the company's affairs.

Innes came accidentally to hear of the terms of the new contract and protested to the Flotilla Company's general manager, who at once tried to wriggle out of it. It cost Finlay much time and effort to hold him to its terms. "I feel more vexed and sore than for years. Busted diplomacy!" he complained to Fleming. He was, therefore, more determined than ever to build a pipeline and by-pass the river for oil transport to Rangoon.

The idea was not new in Burma. As early as 1874 the Twinzas had constructed one at Yenangyaung from bamboos, lacquered inside and resting on wooden stages, to run the oil down the gradual slope to the river bank. It had to be abandoned because of excessive leakages. Since the early 1890s there had been a small metal pipeline along the few miles from the fields to the landing stage on the Irrawaddy. Yet in America since 1865 and in Russia since 1879 oil had been pumped over very considerable distances through pipelines.

Finlay was so keen on the idea that he gave up a good part of his summer holiday in 1901 to an intensive study of all the facts and figures. On his return from holiday he had an on-the-spot survey carried out by an American team. It was completed in the autumn of 1902 and was most encouraging. A 6-inch pipeline some 300 miles long could be laid with few technical problems since practically the whole route lay at about river level. Its capacity would be 1,800,000 barrels a year, or twice the refineries' current throughput, and the overall cost would be 6¾ annas (3p) a barrel, almost half the rate currently being charged. The company would thus save about £40,000 a year and recoup in seven years the total estimated outlay of £275,000. Unfortunately, financial and other difficulties were to delay the pipeline's completion until 1908, as will be related in Chapter VIII.

By 1902 the company's finances were sound enough: so sound, in fact, as to cause problems. Gross profits earned in Rangoon averaged nearly £150,000 a year in 1900–1 compared with £23,500 on average for 1891–4. Profits per barrel refined had kept up well, at about 22p. Asset values had therefore, over time, become unrealistically low. The directors had maintained a highly conservative system of accounting, charging all fields expenditure to income and building up reserve funds to a total of £174,000 by the end of 1901. Since the issued ordinary shares were correspondingly few, totalling no more than £200,000 nominal value, rates of dividend were embarrassingly high; they were (including bonuses) 25 per cent for 1899 and 1901, and no less than 50 per cent for 1900. Therefore shares were changing hands at a considerable premium over their nominal value.

One who expressed himself pungently on this matter was C. W. Wallace, the senior partner in R. G. Shaw & Co., whose Calcutta house, Shaw Wallace & Co., was helping to boost Burmah Oil's kerosene sales in India.[1] He suggested creating a more realistic capital structure by converting the reserves into bonus ordinary shares, as Shell had recently done.

With large searching eyes, tight lips, and high starched collar completely hiding his neck, Wallace kept everyone up to the mark from the humblest office boy to his senior colleagues. Having been born in Calcutta, the son of an East India merchant, he knew Indian mercantile affairs inside out. Finlay was greatly dependent on Wallace's advice, initially over marketing in India and later over policy matters in general. Their offices, at 123 and 88 Bishopsgate respectively, were close enough for the one to be frequently stepping across to see the other.

Wallace's authoritative grasp of financial matters had been acquired in a tough school. Some years earlier, in his mid-twenties, he had helped to put in order the chaotic accounts of the moribund firm which he had then re-formed as Shaw Wallace & Co. Perhaps other firms that were short of managerial talent in those days were equally fortunate to find comfortably off merchants with time on their hands; or perhaps the manner of Wallace's gradual absorption into petroleum affairs was unique in British commercial annals. In any case, Finlay's instincts in cultivating him were sound. Wallace proved to be one of the most outstanding thinkers in the company, alongside Finlay himself and R. I. Watson, who was to become managing director in the 1920s.

Before Finlay had come to any decision about the capital structure, the company's London solicitors brought matters to a head by raising certain objections in April 1902 to a board decision to issue further ordinary shares up to the authorized limit of £350,000. The only remedy was to dissolve the existing company and form a new one with a more appropriate share structure and asset values.

The board engaged as a consultant (Sir) Francis Palmer, a leading authority on company law. Finlay, with Fleming's aid, had to draft detailed proposals and then have them vetted by Palmer and other experts. On 15 May 1902, the initial company was wound up voluntarily and a new company was registered under the same name. The liquidator was John Cargill, the chairman's second son, then aged 35.[2] He had been with the company since its inception, spending three years with Finlay Fleming & Co. in Rangoon before returning to the Glasgow office. The company's day-to-day operations were uninterrupted by this structural change.

The authorized capital was raised from £500,000 to £1,500,000. Existing ordinary shareholders, then holding capital of £200,000, received four new shares for every one they already held, equivalent to the share premium in the market.

Preference shareholders, whose holding totalled £106,670, to their chagrin received no premium when exchanging their shares since the rate of interest was fixed. On the assets side land, building, machinery, drilling plant, tankers and tank installations were all written up to their current valuation.

Besides creating a more appropriate financial framework, the reconstruction permitted future funds to be raised with the least possible difficulty, although only modest sums were initially required. The top management was strengthened by the appointment to the board of John Cargill and C. W. Wallace. The two little guessed that within a year they would have to carry virtually the full burden of running the company.

In the years of consolidation, 1901–3, the Burmah Oil Company was beginning to look more like a modern corporation. In consequence Finlay was more than ever exercised about its standing in the oil world generally. The prime adversary was, as ever, Standard Oil, with which relations had for some time been deteriorating. The jute batching oil agreement still held, but Finlay was convinced that the Americans would welcome the opportunity of snuffing out Burmah Oil. So, when he was informed officially in July 1901 that Lord Curzon, the Viceroy of India, would be visiting Burma that winter and intended to view the oilfields, he saw the visit as a "splendid opportunity" to plead the company's cause in the highest quarters, thereby hoping to gain government protection against the foreigners. The secretary to the viceroy had indicated that a prominent representative of the company ought to be in attendance, and he resolved to go himself.

Curzon's declared economic policy was to give every encouragement to the industries of the empire and, having visited oilfields and refineries in both Russia and the United States, he knew something about the oil business. With almost feverish excitement Finlay confided to Glasgow that a personal meeting with Curzon could therefore be the turning point of the company. "We may make it the Standard Oil Company of the east or we make it a one-horse show," he wrote. "Curzon's confidence has to be gained on the spot."

He duly made his visit; yet his encounter with the viceroy that December fell far short of his expectations. He powerfully stressed the threat from Standard Oil, and was gratified to be given every assurance that the Americans would not be allowed a footing in Burma. In that case, Finlay asserted, Burmah Oil should be made into a powerful asset in the province by tariff protection and liberal treatment over concessions.

To be sure, the existing tariff on kerosene with its 14–20 per cent advantage was some deterrent to importers, but Burmah Oil's low-quality brand was incapable of challenging the superior variety from the United States. No doubt Finlay was looking forward to the time when his company would be producing far more kerosene and of a higher quality; then, so he would reason, a really swingeing tariff increase – say, up to 50 per cent – would be appropriate. However, the India Office had over the years favoured extinguishing, not widening, the tariff advantage. In these circumstances, Curzon did not comment directly on either suggestion during his visit but, just before leaving Rangoon, he sent a private note to Finlay stating that he had asked the government of Burma to submit its views to the top executive body in India, the viceroy's council. The result was that, before the end of the year, Burmah Oil's concession limit was raised from 50 to 150 square miles.

What Finlay could not have discussed with the viceroy was the possibility of foreign rivals scaling the tariff barriers by being granted concessions in Burma itself. In January 1902 the directors of Standard Oil in New York abruptly decided to move into Burma.[3] They applied for concessions in the name of a subsidiary, the Colonial Oil Company, and at once arranged for rig builders and drillers with all necessary equipment to be despatched from the United States. Their Calcutta agent, too, was instructed to make preparations for erecting a refinery in Rangoon. The government of Burma blocked these moves by rejecting the application; the Standard Oil people thereupon sought to buy leases from the Twinzas on unusually generous terms. The government took powers to ban foreigners from acquiring wells, whether through leases or by outright purchase, and forced the Americans' refinery proposal to be dropped as well.

Finlay and his colleagues greeted these resolute countermeasures with much relief. Yet they could not be entirely sure that the local government would maintain their stand once the inevitable comeback from New York occurred.[4] Standard Oil duly tried, but failed, to secure concessions through its British-registered affiliate, the Anglo–American Oil Company. It then sent over its most powerful troubleshooter, William Libby. When reporting the news to Glasgow Finlay portrayed him as that company's "ambassador to all parts of the world", conducting as he went "delicate and difficult negotiations requiring a representative of the highest authority. His work is always done, I am told," continued Finlay, "with the delicacy and urbanity of a successful ambassador."

When Libby passed through London that June on his way to India, he called on Redwood, who learnt that he intended to visit Burma as well and therefore advised Finlay to offer a letter of introduction addressed to Finlay Fleming & Co. At least it would mean that Libby would be conducted round the fields by company employees rather than "sneaking round and dining with the [Twinzas] head men unescorted". Finlay provided the letter of introduction, but privately warned Glasgow to be prepared for almost anything. Libby was known to be making for Simla, the hot-weather capital of India, to "do the diplomatic with Lord Curzon after having done the needful here in London in high places". Finlay's alert to Rangoon was expressed even more strongly. "There is no doubt about it that there is danger – grave danger – about when such a man as Libby is sent out at the worst season of the year, post haste from New York, to interview Lord Curzon."

Notwithstanding Libby's protests in Simla and those to Whitehall from the American ambassador in London, the authorities remained steadfast. When a parliamentary question in July elicited the answer that Whitehall was fully behind the local governments in rejecting the application, Finlay with some jubilation admitted, "We are and have been more successful in this matter than I had dared to hope for."

It was extremely mortifying that at that very moment Burmah Oil happened to fall from grace with those same local governments, which chose – three years late – to censure its absorption of Minbu Oil. Finlay, with unusual crassness, had sought to put officials off the scent by adding ten new names to the list of Minbu Oil shareholders and replacing the existing company documents with others differently phrased. However, the only official sanction was a decree that Minbu Oil would receive no fresh concessions. That did not worry Finlay, as long as Burmah Oil itself was not to be debarred from future ones.

Meanwhile Libby had reached Rangoon; he did not wait to present the letter of introduction but at once went up-country on his own. When word came back that he was negotiating personally with the Twinzas in Mandalay, Finlay Fleming & Co. protested to the lieutenant-governor; yet officials in Rangoon now seemed reluctant to use their legal powers to veto foreigners' purchases of well sites. Finlay hurriedly arranged for well sites to be bought up wholesale so as to deny them to the Americans. He had some minor coups over this, but he could not stop brooding over what Libby and his associates were up to. "A lot of it is bluff," he surmised, "and they are simply twisting our nose," draining into well purchases the company's cash flow from Rangoon and forcing it to raise money in Britain. Indeed, the lack of ready funds was one reason for holding back, for the moment, over the planned pipeline.

The resultant boom in well prices sent the average price of a well site up thirty-fold from under £7 in 1895 to £200 in 1906. It was then to soar another twenty-fold, to £4,000, between 1906 and 1908. Yet despite all the expense of buying up those wells, Finlay felt thankful that he had gone some way towards repulsing Standard Oil, a corporation he regarded with distaste as "a heartless, soulless octopus", bent on "introducing its hateful and destructive monopolistic system into Burma". A further bonus was that in its reconstructed form Burmah Oil was by then reaching that position of strength Finlay had aspired to just eighteen months previously. "We are attaining a point when we can afford to rest on our oars for a bit," he informed Glasgow.

Standard Oil had not yet conceded defeat, however. When in December one of its London-based executives, J. Usmar, came to see Finlay – ostensibly to discuss paraffin wax contracts for 1903 – he made a far more wide-ranging proposal. "Why don't we combine?" he asked, to which Finlay replied artlessly, "We are simple people and only want to be left alone." Usmar then offered to buy a block of the company's unissued capital, whereupon Finlay rather mischievously referred to Shell as a "nice subject for a swallow"; it was in fact less than a year since Samuel had rejected the very attractive bid from Standard Oil, mentioned in Chapter IV, which would indeed have gobbled up Shell completely. Usmar responded with some highly unflattering remarks about Shell and its management. Apart from having to listen to a "diatribe" from the disappointed Libby on his return empty-handed to New York from India in March 1903, Finlay was not to be troubled again directly by Standard Oil.

The two other rival companies, Royal Dutch and Shell, never gave Finlay the same sense of foreboding that haunted him over Standard Oil. "There will, without doubt, in the near future be an attempt at amalgamation in the east, and it is clearly our bounden duty to appear as important as possible," he wrote to Glasgow in August 1902. Indeed, Royal Dutch and Shell had a few months earlier agreed with the Paris Rothschilds – who had Russian oil interests – to merge all their oil marketing agencies in the east under the name of the Asiatic Petroleum Company Ltd.

The architect of the Asiatic agreement had been Henri Deterding, a director of Royal Dutch who had resided in Britain since 1901.[5] Undoubtedly he influenced the course of Burmah Oil's history more than anyone outside the company's own ranks. In his mid-thirties at the time, and thus nearly twenty years younger than Finlay, he possessed exceptional energy and foresight combined with a rare gift for reducing complex problems to their barest essentials. In his case, the hackneyed adjective Napoleonic was not inappropriate. At that time, he had an absolute say in everything he handled, so that all his associates were largely his mouthpieces. By a strange coincidence he, like Wallace, had first made his name by reducing to order some muddled accounts, in his case of a small branch of the first firm he had worked for.

His personality has been much commented on. He was mercurial and strong-minded, with an element of mesmerism in his powerful eyes, which dominated an expressive face. The Burmah Oil directors may have had their own moments of temperament, but basically were among the few immovable objects against which Deterding projected his normally irresistible force. He was fond of a fast and ostentatious life-style that was quite alien to the Scots, who enjoyed their comforts but believed in low-key existences. If Scots are commonly portrayed as frugal and mean, the Dutch have a reputation for being grasping in matters of commerce, and Deterding's biographer, Professor Gerretson, has perhaps helped to confirm this impression by showing him as being at heart a Dutch merchant, with a keen sense for making money.

Had Deterding used his talents merely to rake in the cash, then he would never have been the great man he undoubtedly was. He had a large vision, and as soon as the formation of Asiatic gave him control of most oil marketing in the eastern hemisphere outside Russia, he was able to devise his policy on a global, or at least a hemispherical, scale. His basic ideas were typically simple: above all, an intelligent regulation of supply, not so much to force prices up to monopoly levels as to ensure the production of what customers actually needed.

He also believed strongly in allocating markets to the nearest producer, who should then be able to benefit from the lowest transportation costs. He found it crazy for the Americans to bring kerosene halfway across the world when it was available from nearer sources of supply. Since his favourite saying (probably translated from the Dutch national motto) was "pulling together gives strength", he could be expected to apply powerful pressure to bring outside companies such as Burmah Oil into the combine.

Indeed, having visited Rangoon during an extensive tour of the far east in 1896, he had recognized that Burmah Oil was one of the local producers which would eventually become a thorn in his side. Yet friction had not arisen as long as Royal Dutch had no kerosene to spare for India; but now that wider marketing interests were involved, the proposed Asiatic grouping was bound to bear heavily on Burmah Oil's future plans. As very little information had yet reached the outside world, the Burmah Oil directors could not know whether it was to be merely a cartel designed to shore up prices or whether it had a broader aim of becoming the dominant oil power of the eastern hemisphere. In the latter case it would give no quarter to outsiders, who would be forced into the cartel or out of business altogether.

Where did Standard Oil fit in? That company had already agreed with the Asiatic directors to withdraw partly from India in return for greater freedom to sell kerosene in Europe, thereby economizing in transport costs for all concerned. According to a very important letter written by Deterding to the chairman of Royal Dutch in Holland at the time, Standard Oil's directors welcomed the creation of Asiatic and the consequent tying up of

the eastern market, since it would save the expense and trouble
of having to organize the market themselves. It seems to have
been their prompting that led Deterding to open negotiations
with Burmah Oil, hinting that if the company refused to join
Asiatic, it would be coerced by further efforts to obtain concessions in Burma.[6]

Finlay had hitherto played hard to get with his two rivals.
Despite earlier intentions to live and let live, Samuel had twice
during 1902 sought to induce him to join a kerosene pool that
would fix market shares in India. His somewhat bullying
manner had not gone down well with Finlay, who had maintained his refusal to consider restricting production simply to
allow more foreign oil into India. Deterding's new approach
was more sophisticated. By January 1903, as managing director-designate of Asiatic, he was genuinely anxious about the
uncontrolled rise in Burmah Oil's kerosene deliveries to India
as the extensions to the Syriam refinery came successively on
stream. He therefore went to see Finlay and Wallace and
requested precise figures of expected future shipments to Calcutta. He also proposed a limit on the share of the market he felt
Burmah Oil should enjoy: namely 45 per cent of total kerosene
consumption in Calcutta.

Finlay, while readily agreeing to provide the figures, opposed
the idea of a restricted market share as strongly as he had done
to Samuel. "We are," he explained, "under direct obligation to
government not in any way to shut down or modify the
development and output of the Burma oilfields." He and
Wallace had already agreed among themselves that the only
possible way of avoiding a collision with Asiatic was some
scheme to divert part of their own increasing kerosene supplies
from Calcutta to markets outside India. As he later reported to
Glasgow:

> Having reached this point [of an impasse], we drew from Mr.
> Deterding exactly what we wanted, i.e. that his company should
> buy from us any surplus not to be sent to Calcutta, at a price, and to
> be diverted by them in their own steamers.
>
> He raised all sorts of difficulties in regard to this suggestion, but we
> took up rather a severe attitude with him, and brought him down to
> the anvil of commonsense.

But what did he mean by "at a price"? Under pressure, Deterding indicated that he would be willing to buy surplus kerosene at the equivalent of 5d (2p) a gallon, a figure that even the pessimistic John Cargill conceded would leave the company a very handsome profit.

Finlay still feared hostile government reactions should the company be forced into some such accord with Asiatic. Yet by the end of 1903 the sixth extension of Syriam would be completed and the flow of kerosene would be high enough to leave no room for the Asiatic Petroleum Company in India. He therefore suggested to Glasgow, "Would it not be better in the company's interests, seeing that we can produce oil so cheaply and will be able to produce it still more cheaply, to say 'No thanks, gentlemen. We have nothing to offer you. We will pursue our course independently'?"

Deterding sought to force his hand by opening a price war and by sharply reducing Asiatic's kerosene prices in Calcutta. Wallace, already impatient that the company was going "too far in hand" with Asiatic, was all for following these prices down immediately; that would also ease the problem likely to arise when increased deliveries came from Syriam a few months later. Finlay admitted to being on the horns of a cruel dilemma. To cut prices would reduce the company's cash flow at an inopportune time. On the other hand, to enter into an "unholy combination" with Asiatic might prejudice vital government backing and give Standard Oil the one thing it needed to get into Burma.

If he still had a deep-seated fear of the Americans, Finlay seemed nevertheless to underestimate the threat that might come from Deterding. The Asiatic directors, he wrote with more than a touch of complacency, "must know they cannot beat us. . . . They are in a funk and their attitude towards us is timorous. Before it was arrogant." Whether in a funk or not, Deterding was biding his time, keeping in touch with Finlay both directly and through an intermediary. Yet by the end of March 1903, a war of attrition between Burmah Oil and Asiatic was well under way.

All these strains and exertions must have been slowly wearing down Finlay's health. For the past seventeen years or so he had been carrying the main burden of the company's day-to-day affairs almost single-handed. In the summer of 1901, he had been far from well, his usual buoyant mood giving way to uncharacteristic jumpiness, so that self-pitying phrases such as "I'm tired and late" crept into his correspondence with Glasgow. The trip to Rangoon that winter, despite all that had to be fitted in, had done him some good; then a return to the rigours of an English January had brought on a severe chill. That chill had undoubtedly been aggravated by the inadequate heating in his large Ealing house. He had married well: his wife Jane was the daughter of a Glasgow wine-merchant, but in spite of their affluence – he was worth £125,000 after the company's reconstruction in 1902 – they lived frugally, with only a few servants and the lady of the house taking a pride in polishing the furniture herself.

Finlay's executive burden was increased when his chief mentor, David Cargill, began to fail and early in 1903 attended a board meeting for the last time. At the age of 76, Cargill was confined by paralysis to a wheelchair at his country home of Carruth House at Bridge of Weir, some 15 miles from Glasgow. His son, John Cargill, therefore had to assume responsibility for the oil business as well as the family trading activities. Although he had gained some confidence from acting as liquidator in 1902, he was relatively inexperienced and by character averse from taking decisions. He could not be the steadying influence on Finlay that his father had been. Fleming could be relied on for down-to-earth advice, but did not reckon to put himself out for anyone.

Henry Ford, the car-maker, once said that time given to the study of the competition was time lost for one's own business.[7] If so, the time Finlay could spare for constructive work was all too limited. Although he could have done without them, a series of tedious negotiations took up much of his energies from the summer of 1902 onwards. These were with Moore and Mower over the Rangoon Oil Company. Finlay offered either to purchase its crude oil at the same price as the Twinzas received, or, alternatively, to provide funds for the creation of a greatly enlarged company in which Burmah Oil would hold the controlling share. The Rangoon Oil people rejected the second proposal and demanded a price well over the odds, claiming that Standard Oil and other bidders were after them as well, and that the lieutenant-governor was anxious to grant them

further concessions to undermine Burmah Oil's powerful position. Faced with a stalemate, Finlay decided on a swift coup to purchase Rangoon Oil shares until he had a majority control and outface the authorities should they take umbrage at his absorption of a rival.

For once he seemed to lose his touch; although Finlay Fleming & Co. strove to buy up all Rangoon Oil shares regardless of cost, it failed to secure even a quarter of all the shares. Moreover, Finlay no longer had the steadfast support from Glasgow he had enjoyed during David Cargill's active years: the son, John Cargill, was desperately worried that the quarrel with Rangoon Oil was costing the company £40,000 a year, a quarter of the year's net profits, and was ready to give in to Rangoon Oil's demands. By Christmas the matter was still unresolved. Finlay went through agonies over the holidays, at first accepting and then rejecting the terms. That final rejection came as a great relief to Boverton Redwood, who condemned the terms as absolutely unreasonable and merely an encouragement to rivals to spring up with the sole intention of being bought out.

Finlay was nevertheless to pay dearly for cold-shouldering Rangoon Oil. Early in January 1903 he received the calamitous news that Standard Oil had offered to buy the whole of that company's crude production at the higher price demanded, and that the lieutenant-governor had approved the deal. For a short while he was shocked into a state of inertia, convinced that the latest outrageous move was "a mere blind enabling the Standard Oil Company to enter by the back door". He now bitterly regretted having turned down Rangoon Oil's previous offer.

The atmosphere of tension in the overcrowded office at 123 Bishopsgate must have been perilously high. Physically, Finlay had filled out since the Rangoon days when he was a pale young assistant. He now had a good presence, with a military type moustache, and his appearance suggested a person of some consequence. Yet the stream of cables from Rangoon and telegrams from Glasgow, the telephone with its maddening delays and even more maddening lack of clarity, the necessity of absorbing every word of the lengthy correspondence and replying decisively by the weekly mail day – all these imposed a pressure on Finlay as remorseless as any machine of the day could have done on the operatives. He had only clerks to help him. Had he been able to delegate to subordinates, he could have saved himself much stress, but he did not trust anyone to do his work as well as he felt he did it himself. Whether or not Standard Oil had made a genuine offer, Rangoon Oil's tactics had both confused and panicked him.

A day or so later he rallied, and accepted the asking price for the crude oil. No more was heard of Standard Oil's bid and in February, an agreement was signed, to last five years. As with all manic-depressives, he veered dangerously fast between extremes of moods. "We may well congratulate ourselves upon having brought the negotiations to a successful issue," he wrote pragmatically to John Cargill. What Boverton Redwood thought of the climb-down, he kept to himself.

Even so, the disposal of that tricky question appeared to bring no lasting relief to Finlay's mind, but merely reawakened his fears over Standard Oil. How much he wished "the papers would drop crowing" over the Americans' rebuff in Burma, he told John Cargill, since much more gleeful comment would "only make the Standard Oil Company angry and lead them to make reprisals". It was more likely than ever that Standard Oil would be seeking alternative sources of crude in the east, as by various mischances it had been running very short of crude at home.

Meanwhile, problem seemed to pile upon problem as the price war with Asiatic was still depressing the company's income earned in Rangoon. Yet Finlay was becoming convinced that "the most ticklish thing" he now had to handle was the issue of the pipeline. Innes and the other directors of the Irrawaddy Flotilla Company were exhausting every power at their disposal to prevent Burmah Oil from ever abandoning the transportation of crude by river. When Standard Oil had earlier enquired about the rates that company would charge, its directors had happily quoted the same rates as they charged Burmah Oil. Finlay therefore inferred that they would be only too delighted to see Standard Oil established in Burma, as they feared Burmah Oil had become "too big and strong for them".

In April 1903, Innes yet again raised the pipeline question. Finlay bluntly told him that the threat from Standard Oil remained so real that Burmah Oil could no longer burden itself with an unfavourable long-term freight contract. If the Irrawaddy Flotilla Company refused to handle the company's oil after the existing contract ran out at the end of 1904, he would accept a temporary interruption to output as he was resolved to complete the pipeline. In any case, the existing freight charges were still too high, and he quoted comparative figures of the rates charged for the 500 miles from Baku to Astrakhan in Russia to show that he was currently paying twice as much for transport over half the distance.

Innes responded constructively by agreeing to extend the contract on existing terms until the end of 1905 and on a yearly basis thereafter. Then on 15 April Wallace, who had increasingly become Finlay's adviser and right-hand man, was found to have an abscess on his liver and was taken off for immediate surgery. He would therefore be out of action for some months to come. "I greatly regret this, as I relied on Mr. Wallace upon many points where his expertise was very great and of inestimable value to me," Finlay observed mildly to Glasgow. Yet that event turned out to be the culminating blow to a sorely overstretched mind. He travelled north to the new company's first annual general meeting, which was chaired by Fleming in David Cargill's absence. Although managing director, he was not called on to speak.

He must already have suffered the onset of a breakdown; yet what precisely happened in the next few months can only be conjectured. His son, Campbell, then an assistant in Finlay Fleming & Co., had recently come on leave from Rangoon with his wife and baby. They stayed at Ealing, and Kirkman Finlay was noticed to be talking gaily and amusingly at table, relishing his son's company. Then Jane Finlay told her daughter-in-law that they were all making too much work for the servants, and they moved on to some relatives in Ireland. From there the son was hurriedly recalled when his father's mental state became apparent, and arranged for him to be taken, under the care of two attendants, to a hotel at Newquay. There, on 27 June, he gave his attendants the slip. His body was found by a fisherman off Harlyn Bay, and the coroner for Cornwall brought in the verdict: "suicide by throwing himself over the cliff whilst insane."[8]

In an age when the workings of the human mind, whether in health or sickness, were far less well understood than they are today, the directors were not over-anxious to publicize this shocking tragedy. His death certificate ambiguously described him as an oil merchant. Not until thirty-four years had gone by, and after the death of his son Campbell, did John Cargill at an annual meeting pay the tribute that Kirkman Finlay so richly deserved, and attributed his sudden demise to many years of overwork and overstrain.[9]

Cargill himself had every reason to remember the dreadful few days after the disappearance was reported. He had hurried to London and, as soon as the body was discovered, had to break the news to the widow. She collapsed, and within a few hours was dead of heart failure. The couple, aged 55 and 53 respectively, were given a joint funeral the following Saturday at the unusually late hour of 4 pm and buried together in Highgate cemetery. A guarded notice that the company inserted in the "deaths" column of *The Times* on the actual day of the funeral concluded: "Please accept this, the only intimation and invitation."[10] It is not known how many mourners were present.

By an irony the dead man would have been the first to appreciate, the authorities in Whitehall had already taken the initial steps – without the company's knowledge – towards making it a supplier of fuel oil to the Admiralty. Whitehall's decision was in due course to guide the company towards areas of enterprise that could scarcely have been imagined in 1903.

Finlay had been wrong a year or two back to see the alternatives for Burmah Oil as remaining a "one-horse show" or becoming "the Standard Oil Company of the east". There was a third path – which he might possibly have deplored – of being a medium-sized company with interests that stretched well beyond Burma. It was along that path that the company was about to be directed.

Notes

1 Townend, *History of Shaw Wallace*, p. 2. The whole of ch. I gives valuable evidence of Wallace's early connections with the company.
2 For John Cargill see "The Bailie", "Men You Know, No. 15", *Glasgow Weekly Herald*, 9 July 1932, and obituary in *Journal of Institute of Petroleum*. 40, 1954, p. 88.
3 R. W. Hidy and M. E. Hidy, *History of Standard Oil Company (New Jersey) I*

Pioneering in Big Business 1882–1911 (1955), pp. 499 ff. for a discussion of this attempt to enter Burma.

4 For 1902 correspondence arising out of the official refusal of Standard Oil's application see IOL L/E/7/414, 426/99, especially W. H. Libby to Secretary of Government of India, 14 July 1902.

5 A fully documented life and times of Deterding (1866–1939) would be of great value. A personal biography, written by a non-oil man, Paul Hendrix, is due to be published shortly, in the absence of which Deterding's own "memories and reflections", *An International Oilman* (1934), should be consulted. The English translation of Dr F. C. Gerretson's *History of the Royal Dutch* in four volumes (1953–7) runs to 1914. A new Dutch edition with footnotes, in five volumes, was published in 1971–3; volumes 4 and 5 have extra material up to 1923. It must be understood that Gerretson's work, in contrast with dry-as-dust narratives such perhaps as the present one, is really a saga with a hero (namely Deterding, for whom it was written as a surprise present) and a suitably eloquent final section which portrays the scene in Royal Dutch head office when war broke out in 1914. Many of the references to Burmah Oil are inaccurate (see ch. VI note 15.).

6 Henriques, *Marcus Samuel*, p. 453.

7 L. Hannah, *The Rise of the Corporate Economy* (1976), p. 80n.

8 *West Briton*, 2 July 1903.

9 1937 AGM, *Glasgow Herald*, 5 June 1937.

10 *The Times*, 4 July 1903.

CHAPTER VI

The Admiralty and Asiatic
1903–5

Kirkman Finlay's untimely death left a void in the Burmah Oil Company at a critical stage in its development. The sheer weight of intractable problems, which may have precipitated his final breakdown, did not vanish overnight. Apart from those to do with finance, the company still sought further concessions from the government of Burma; Standard Oil continued to make persistent efforts to gain a foothold in the province; Shell and Royal Dutch, already forming the powerful "Asiatic" cartel, were poised to inflict great damage on the company's sales in India; competing producers within Burma, from Rangoon Oil to the Twinzas, were as active as ever; and the disagreements with the Irrawaddy Flotilla Company on river shipments and a future pipeline were yet to be finally resolved.

Thus Finlay's place in the London office might have been expected to be filled without delay. In fact, when Wallace declined the managing directorship because of his poor health, both that post and the London desk were kept vacant. Instead, John Cargill preferred to strengthen the Glasgow office – already depleted as a result of his father's terminal illness – by recalling R. W. Adamson, the general manager in Rangoon. As a director in Glasgow, Adamson dealt with an immense amount of paperwork, chain-smoking Burmese cheroots until he was almost invisible through the fumes. Although not an imaginative man, he had twenty-five years of experience in Burma and was well endowed with sturdy commonsense.

C. K. Finlay returned to Rangoon as general manager there. With an intelligent, rather John Buchanish face and a deep but sceptical interest in religion and philosophy, he had studied for the Indian Civil Service and had been called to the bar. Those meeting him for the first time might have thought him a Hamlet-like character. In fact, his appointed task as well as his inclination was to wake up the rather somnolent Finlay Fleming & Co. As John Cargill stated a few years later, breaking unconsciously into rhyme, "Mr Finlay as you know is full of energy and go, and likes his assistants of the same stamp." He remained in Rangoon until he came home for good in 1912.

Cargill himself and the company secretary, James Hamilton, concentrated on their specialized tasks and travelled to London only when any major question made this necessary. The burden on the 36-year-old John Cargill of carrying the whole company and the related trading activities must have been formidable. Almost painfully diffident, he later admitted that at times in that period he had felt overwhelmed by his responsibilities, but drew strength above all from the help and encouragement of Boverton Redwood.[1]

As it happened, the two and a half years that followed Finlay's death confronted the company with several entirely novel problems. The first was a pressing need for economy. John Cargill, when he took charge, was resolved that something must be done about waste. Above all, he insisted on Finlay Fleming & Co. and through them the refinery and drilling people being made more cost conscious. In December 1904 he declared, "There is no doubt that in many directions a considerable amount of extravagance had crept into the working of our business." It was some years before the true magnitude of that extravagance was understood.

The second of the company's novel problems was that of learning to live with a long-term contract for the supply of fuel oil to the Admiralty, signed in November 1905,[2] that restricted freedom of action, as did a kerosene agreement concluded with Asiatic about the same time. Hence change was being imposed on the company just at a time when the men most capable of carrying it through were no longer available.

For several decades, the Admiralty had been interested in oil propulsion, despite plentiful and cheap domestic supplies of coal and the massive technical and other difficulties that any change-over to oil would involve. As early as the 1880s, its warships started to use a little oil for spraying on coal so as to speed up the raising of steam. Once several types of workable marine oil-boilers had been developed, the Admiralty began serious experiments to see if any might be suitable for propulsion itself.

The potential advantages of oil for naval efficiency in peace and war were enormous. It would add half as much again to the effectiveness of any fleet through, for instance, reducing by a quarter the number of stokers and others in the boiler rooms. A coal-burning vessel running short of fuel had to return to harbour, where virtually the whole ship's company endured the grimy and exhausting task of coaling ship. That problem was becoming critical as warships were equipped with ever larger engines to develop far higher speeds. Oil, on the other hand, needed only a small working party to connect up the hoses, after which the fuel was simply pumped in, even at sea if necessary.

Those were the powerful arguments of John Fisher, a full admiral since 1901, who was a rumbustious and very active campaigner for oil. Yet despite his reputation as an "oil maniac" and persistent lobbying of political and naval superiors alike, he was largely campaigning from the outside. Although good at firing off memoranda or letters, he produced fewer definitive results than did the more painstaking efforts of less publicity-conscious contemporaries, notably those in the engineer-in-chief's department of the Admiralty.

There the first experiments had begun in 1898–9 and, as a result, the first petrol-driven British submarines were ordered in 1901. Not until four years later were oil-burning coastal destroyers constructed, but a standing committee in the Admiralty, the Oil Fuel Committee, was set up in 1902 to examine the availability of oil and take all possible steps to bring additional oil supplies under British control. From 1903 onwards the chairman was the parliamentary secretary, E. G. Pretyman. Other members included Gordon Miller, Director of Navy Contracts, and Boverton Redwood.

In the early months of 1903, Miller and Redwood gave important evidence before the Royal Commission on Coal Supplies, set up as a result of fears, later found to be much exaggerated, of Britain's coal reserves running out.[3] This was alarming to the Royal Navy, which depended almost entirely on coal. The commission therefore enquired into possible

1. David Sime Cargill
(1826–1904), founder and first
chairman.

2. Sir John T. Cargill, Bart.
(1867–1954) chairman
1904–1943.

3. The hand dug wells at Yenangyaung, from F. Noetling's *The Occurrence of Petroleum in Burma* (1898).

4. The first well at Digboi (Assam 1889, taken over when Burmah Oil acquired the Assam Oil Company in 1921.

B.O.C. No.1
STARTED 1ST NOVEMBER 1886.
FINISHED 5TH MARCH 1889.
DEPTH 727 FEET
INITIAL YIELD 500 G.
DRILLER L. HICKSON

5. The site of the Burmah Oil Company's first machine drilled well, completed at Yenangyaung in 1889.

6. Sir Boverton Redwood (1846–1919), consultant to the company from 1893 to his death. From a cartoon by Spy. (Reproduced by courtesy of the Institute of Petroleum)

7. The first well at Chiah Sourkh, Persia, drilled by the Concessions Syndicate Limited, 1905.
(A BP photograph)

8. The difficulties of transporting oil drilling equipment in Persia, c. 1905.

alternative fuels, but oil had the overwhelming disadvantage of having to be imported. The Scottish shale industry produced no satisfactory fuel oil, and sources within the empire were not much more encouraging. Canada, with a limited and apparently dwindling output, certainly had none to spare for export. That left Burma, which contributed 1 per cent of the world's production.

The Royal Commission was therefore anxious to know something about Burmah Oil as the province's main producer. Redwood spoke up for the company, and stressed its conspicuously successful enterprise, vigorous development, and potential for considerable future expansion thanks to systematic and energetic planning for growth. He did not add that the plans were not as encouraging as had been hoped, thus precluding the possibilities of really massive expansion, and he probably laid it on a little thick when he added, "It is satisfactory to find that this spirited policy is receiving the well-merited support of the government." Yet he had made sure that the company was now firmly within the Admiralty's sights.

Whether as a result of Redwood's evidence to the Royal Commission or of his prodding on the Oil Fuel Committee, in May 1903 a high official of the Admiralty privately requested from the government of Burma precise information on oil supplies there. He asked no question about fuel oil as such, but the reply corroborated Redwood's evidence so well that at the end of July Gordon Miller wrote officially to Burmah Oil. He asked whether it had fuel oil to offer from its production and, if so, whether samples could be made available for analysis and eventual testing.

Had Kirkman Finlay still been alive, he would have known enough to furnish at least an interim reply within a matter of days. Lacking the basic information, his son – then still in the London office – had to forward the letter to Rangoon. There Adamson and Andrew Campbell drafted a reply that was in due time sent direct to Glasgow. A further three months went by before, in November, the company at last despatched its reply to the Admiralty.

The reply made it clear that the company organized its refining operations so as to maximize the output of kerosene, its most lucrative product, the residuum or *astatki* being used as a fuel either in its steamers or for the boilers in the refineries. It was currently unable to supply fuel oil in any quantity, but could do so once the next set of extensions at Syriam had been completed "at no distant date". However, it continued, secure supplies of crude oil would be vital. Official limitation of the company's concessions to 150 square miles was "scarcely in keeping with the position of the company as the pioneers of the oil industry in Burma" – but, the letter added, "we are hopeful that this limit will be increased later on. . . . This history of the oil industry in India is the history of the Burmah Oil Company and what has been done in the past is a criterion of what is being done now and of what will be done."

Redwood, when the Admiralty asked him to comment on the company's reply, declared that Burmah Oil would be perfectly capable of supplying naval fuel in considerable quantities, being financially sound and already in possession of adequate crude reserves. He advised giving the company inducements, such as further concessions and an increase in the tariff, to extend its operations. At the same time, he criticized the current policy of seeking to encourage rival firms which necessarily lacked Burmah Oil's special knowledge and experience.

Cargill, who replied to a further Admiralty enquiry by professing his readiness to help in any way possible, was by no means anxious to embark on perhaps very far-reaching commitments with the Admiralty at a time when major competitors were once again on the warpath. Only two days after Kirkman Finlay's death, the long-awaited registration of the Asiatic Petroleum Company had taken place in London. In September, Deterding visited Cargill and Wallace to renew his suggestion, last put forward six months earlier – although it now seemed like another age – that Burmah Oil should restrict its kerosene sales to 45 per cent of the Calcutta market and sell the surplus to Asiatic. The two replied that they were only too happy to work amicably with him but would not enter into any restrictive agreement nor jeopardize the planned increase in refinery capacity; Deterding was, however, welcome to their forecasts of future deliveries to mid-1904.

Then in November three Shell directors, including Samuel, came to see Cargill and Wallace while Deterding was on holiday, and proposed an end to the price war in India that

Asiatic had started in January. Apparently for the first time, Asiatic conceded the company's right to a prior market in the subcontinent and did not object to its determination to impose a maximum price on its Victoria brand of kerosene. The level of this maximum price, which was to play a key role in the future, seems to have been chosen by the Burmah Oil directors as one that was both remunerative and an earnest to the government of India that they were not out to sweat customers.[4]

No sooner had Asiatic steadied the market by raising its prices than Standard Oil began to take threatening steps against Burmah Oil. Early in 1904 it looked as though war to the death, such as Kirkman Finlay had dreaded, was under way. Mercifully, the previous year's financial results had been reasonably good, with profits remitted from Rangoon increased to £332,000, or 34p a barrel, compared with 22p in 1902. Refinery throughput was up by 40 per cent compared with 1902, which had helped to bring down unit costs and to offset the lower prices resulting from the rate war. Prudently, the directors put nearly a third of those profits, about £90,000, to general reserve.

Redwood took a fairly relaxed view of Standard Oil's moves, seeing them as merely "a bluff to rope the Burmah Oil Company into a combine", but the directors were taking no chances. In April they instructed Finlay Fleming & Co. to send a memorial, or petition, to the viceroy asking for the tariff to be raised from 14 per cent to a penal 56 per cent of the retail price.[5] It also detailed recent tactics in India of Asiatic and Standard Oil, to prove that they were working "in perfect concert" to drive Burmah Oil out of the subcontinent. In Ceylon, for instance, customers had to pay nearly three times as much for kerosene as in India because the two companies had a monopoly there. Thus if Burmah Oil were knocked out of India, consumers would be £4–5,000,000 a year worse off through higher prices. Standard Oil's attack petered out, with no lasting harm to the company.

In the same month of April 1904, Cargill and Wallace for the first time visited the Admiralty to meet the Director of Navy Contracts, Gordon Miller. They corroborated Redwood's information that the company was capable of producing the required specification of fuel oil, but professed their main anxiety to be a financial one. To supply 50,000 tons (312,500 barrels) a year, the company would need a special plant which would be costly and take at least twelve months to erect.

Cargill did not reveal the bother he was having with Andrew Campbell, the works manager, who doggedly opposed the idea of earmarking the seventh extension of Syriam for fuel oil since it would upset his carefully conceived plans for developing the refinery to maximize kerosene yields. The ever pessimistic Cargill saw as a "relief" the need to modify Campbell's plans: as he wrote to Finlay in Rangoon, he was beginning to view with considerable anxiety the question of marketing an enormously increasing production of kerosene in face of the continuing hostility of Burmah Oil's main rivals.

To Admiralty officials, the directors' obvious wish to help in any way created a highly favourable impression, which clearly assisted Redwood in his regular discussions with those inside and outside the Oil Fuel Committee. When reporting to the directors in May, he passed on Pretyman's view that the government of India was "anxious and determined" and the Admiralty "if possible, even keener" to see the company's interests protected. Yet, he hastened to add, such solicitude might not last indefinitely, since no government could bind its successors any more than it would permit a comprehensive oil monopoly, even an all-British one, in Burma.

That anxiety for the future was heightened when they heard that the Shell group was seeking to make its way into Burma by applying for prospecting licences through its Calcutta agents. It had also sent an expert from Royal Dutch and a Shell geologist to make investigations on the spot. The application was in due course turned down, but it had been a nasty scare.

By then a sample of fuel oil from Rangoon had been tested at Portsmouth and found equal in quality and convenience of use to top-grade Texas fuel. The Admiralty therefore asked for 600 tons, to be used for full-scale seagoing trials. For the first time, Pretyman enquired about the price the company proposed to charge. Cargill and Wallace explained that it would be on a sliding scale, depending on what kerosene was fetching in India at any given time. As the price war in India had driven kerosene prices down almost to the lower level of their scale, fuel oil would have to be that much more expensive. A case for at least doubling the tariff therefore seemed overwhelming.

Pretyman at once objected that tariffs were intended to yield revenue, and that any such increase would merely provoke criticism that the government was acting to "offer a monopoly to the company". Cargill and Wallace, therefore, fell back on pressing for further concessions. At their next meeting, Pretyman did offer them the solace that the government of India had no intention of "letting the company be beaten out of the field", although it was powerless to move until a definite threat arose. Cargill on his side agreed to plan for 15–20,000 tons of fuel oil to be produced in 1905, and promised to help the Admiralty over the chartering of tankers in due course.

David Cargill, although nominally still its chairman, had taken no active part in the company's direction since early in 1903. In May 1904 he died, aged 78, worth £943,000, and was succeeded by his son, John.[6] Despite all that he had done to keep oil operations going in Burma before 1886 and in promoting the company's subsequent growth, David Cargill remains a shadowy figure. No personal letter of his survives, nor can he be positively identified with any specific facets of the company's policy. He was remembered as combining in roughly equal degrees great courage with great caution, and breadth of vision with an unquenchable passion for detail.[7] Such a combination made him a solitary man. Every evening after dinner he would withdraw to his study and work at his papers far into the night, for his "old enemy" of long standing was insomnia.

In his earlier days, he had been a patron of the arts. However, unlike Kirkman Finlay, he seemed always to be too busy to join in much of the family life going on around him. With five children by his first wife and two by the second, whom he married in 1878, he had domestic responsibilities enough. Yet the singlemindedness of an entrepreneur may involve heavy sacrifices in terms of family life; undoubtedly the Cargill boys turned out differently from those in other comfortably off Scottish families.

The eldest son James took the drastic step, for the son of a Church of Scotland elder, of becoming a Roman Catholic and joining the order of St Philip Neri at the Brompton Oratory in London. The third son, David, went into the company and became a director in 1910, but does not seem to have influenced policy much; he married twice, had no children and died of alcoholism. The youngest son never married and devoted himself to amassing works of art; his collection of French impressionist paintings was sold after his death for over £1,000,000. There was not one grandson in the male line to carry on the Cargill name.

As to the second son, John, he became the very model of a paternalistic company chairman actively concerning himself in the welfare of his employees at home and abroad. Very tall, with a normally benign expression and an impressive Roman nose, he must have towered above those around him in almost any company. Yet he, too, was fatally flawed. He later recalled his own neglected upbringing and to the end of his life displayed many of the characteristics of a deprived child: a marked diffidence and acute anxiety about most things, combined with a stubborn will to be right over matters where he claimed expert knowledge. When subordinates were obviously afraid of him, he could be a bully; but he respected those who stood up to him, despite the "battles royal" that often ensued.

His strange, deeply pessimistic character can best be seen in his handling of the annual general meetings. In the early years he dreaded them. As time went on, his habit of qualifying the rather impressive results with despondent warnings of disasters to come grew into one of the longest-running jokes among Glasgow's commercial fraternity. After the 1917 meeting a Rangoon paper rather unkindly portrayed him as a skilled mourner who always addressed his shareholders with a sense of impending doom. This led him the following year to venture the comment – received with huge delight – that if doom befell him and the company, he could safely rely on getting a position with the city's top undertakers "and would thereby be able to meet you at various gatherings during the year, and you would be able to judge whether I am a better chief mourner than chairman of this company."[8] Like many aloof men of the highest integrity, he eventually received some measure of the affection he so obviously craved.

For its geological expertise the company had hitherto relied mainly on the Geological Survey of India: Noetling and Grimes had between them mapped and delineated the three principal fields of Burma. In those days the company's American drillers were entirely responsible for siting wells and, indeed, for developing whole fields. Tradition has it that they did not know exactly where to sink the first well at Singu, and one of the drillers suggested, "Let's put it down where that bird settles." They did so, and the outcome was more fortunate than it deserved to be. Not until August 1904, when the Admiralty was demanding increasing quantities of fuel oil, did the directors accept that they needed their own geologists. C. B. Jacobs, fields manager since Seiple's retirement in 1902, could not guarantee to produce enough oil to satisfy Admiralty requirements, so C. K. Finlay from Rangoon asked for a geologist to discover new sources of crude.

Redwood, therefore, loaned a member of his staff, W. H. Dalton, who went out with two assistants late in 1904. Dalton was horrified to discover that maps were lacking for some of the most productive territory. He did not think much of Jacobs: "an energetic man, but uneducated and narrow", who proved a hindrance while accompanying him and his team. Later, as a consultant, Dalton trained the first permanent geologist, Basil Macrorie, who went to Burma in mid-1905. A year later the company had four or five geologists, who between them made its prospecting far more professional.

Yet the oilfields needed more than a team of geologists. To impose some order there, in 1905 the company appointed a fields agent who would reside on the spot. He was to be an office man from Finlay Fleming & Co. to relieve Jacobs of the job of liaison with the Twinzas. Plainly, this appointment was the thin end of a wedge that would enable the agent to intervene forcibly in achieving economies.

On refining, the technical departments in Glasgow sought to discover and investigate all new processes that might be of value. These did not always go down well with the mulish Campbell in Rangoon, whose usual response to any fresh idea was that his own system of refining happened to be the most satisfactory in the circumstances. His aim was to achieve a delicate balance, whereby kerosene production was maximized but enough residuum produced to fuel the refineries. He did concede that his Victoria kerosene was not very good, even if it was easy to produce.

Naphtha, formerly burnt off as useless, was becoming more valuable. From 1905, Campbell began to produce some petrol, which was in growing demand now that the first motor cars were appearing in the east. Output totalled only 1,000 barrels in 1905 but exceeded 1,400,000 barrels by 1914. Technical progress also enabled him to extract superior quality kerosene, and the Gold Mohur brand was marketed for the first time in 1905.

The beginning of 1905 brought the directors the welcome news that officials in Britain and in the east were taking a more realistic attitude towards Burmah Oil and its problems. In 1900, the under-secretary of state at the India Office had been anxious to sweep away all measures of "artificial protection" through tariffs and the like; five years later he conceded that such an out-and-out free trade policy had become outdated now that there was a new protectionist spirit of "tariff reform" in government thinking and the Admiralty had proclaimed the importance of the Burma oilfields for imperial reasons.[9] In India, Curzon as viceroy had just set up a department of trade and industry and the first head, Sir John Hewett, soon became an ally of the company. He pressed for the proposed fuel oil agreement to be signed as quickly as possible and to cover the largest practicable quantity. That would give the Secretary of State a watertight case, on grounds of national security, for refusing concessions to foreigners.

Yet the directors' anxieties persisted, especially as the company's results for 1904 had demonstrated the costly effects of the long drawn-out price war: profits per barrel had fallen from 24p to 19p. As the only way of paying a reasonable dividend, the company had to put to capital expenditure the £182,000 spent that year on drilling and equipment. Cargill made sure that Miller at the Admiralty understood the difficulties caused by the price war.

As it happened Deterding, who had remained in the background for over a year, decided that the time had come to bring Burmah Oil to heel. He despatched his chief assistant, Robert Waley Cohen,[10] then 28, to the east, ostensibly to discover why kerosene prices and earnings in India, one of Asiatic's major markets, were so low. Cohen also had secret instructions to pave the way for an eventual accord with Burmah Oil.

His biographer has likened Cohen's tactics to those of Standard Oil's John Rockefeller in inflicting a sharp lesson on the chosen prey and then softening him up with gentler treatment before forcing him into an agreement.[11] Cohen's first move in India was to agree with Standard Oil to push up kerosene prices and then drop them sharply in concert and unload the substantial stocks they had both built up. Hoping thereby to cripple Burmah Oil's sales, he would then approach the company's representatives in order to drive home "the folly of their ways". Unfortunately, Deterding rather blunted the impact of this master-plan by decreeing that Asiatic's objective must be to capture the largest possible market share regardless of price.

To the Burmah Oil directors, Cohen's Indian trip seemed likely to stir up further trouble. Then he had a quite unexpected change of heart. Before he left Calcutta for home in February 1905, he called on Greenway and proposed a working arrangement that would leave out Standard Oil entirely. Asiatic would be allowed to market 1,000,000 barrels of kerosene in India and Burmah Oil the remaining 800,000 barrels, say, without having to restrict its production. All the company's surplus would be sold by Asiatic in the far east, the price being slightly less than that of imported Russian oil. Cohen quoted Rs15–16 (£1.00– £1.10) a barrel as low enough to discourage Standard Oil from importing it in bulk.

Greenway passed on the proposals, predictably emphasizing the risk of tying the company to Asiatic in any way: C. K. Finlay in Rangoon cordially agreed. Cargill at once restated what he had said to Samuel more than a year before: no agreement, but a measure of co-operation should the Asiatic directors halt "the insane war of prices", provided that Burmah Oil could still market its entire output. To Redwood he wrote with some assurance, "it is very evident from these overtures how the shoe is beginning to pinch" for Asiatic as a result of the price war.

Then, after time for reflection, he developed an attack of cold feet and Wallace had to bring him round with the cheery remark, "I don't think that there is anything to fash ourselves about there"; by approaching Greenway, Cohen had exposed Asiatic's weak position. Asiatic was free to make contact again whenever it was sick of the long drawn-out price war and meanwhile Burmah Oil was committed to nothing. Undoubtedly, Cohen had discovered on the spot that there was no chance of driving Burmah Oil out of the Indian market and that a negotiated settlement was the only course of action.

As the year 1905 wore on, things seemed to be looking up for

the company. The Admiralty remained as enthusiastic as
before for whatever Burmah Oil could provide in the way of fuel
oil. However, by then the Conservative government under A. J.
Balfour was in a very disordered state and appeared unlikely to
last the year out. The directors were well aware that a new
government might not be so willing to keep foreigners out of the
Burmese oil industry. In any case, when asked formally by the
Admiralty to end the practice of reserving blocks against the
company, the India Office refused, on the grounds that Burmah Oil must be prevented from securing all the best blocks
and thereby acquiring a monopoly of the industry. Yet the
Admiralty at last admitted that the company would derive no
profit from the fuel oil agreement, and wrung some minor
concessions out of the Indian authorities.

Then, quite unexpectedly, the company obtained some mor-
al support from one of the most exalted official bodies in
Edwardian Britain. A few years before, the Committee of
Imperial Defence had been set up under the chairmanship of
the prime minister. In August 1905, its secretary wrote to other
government departments, stating that the oilfields of Burma
were of vast importance. Bearing in mind the enormous capital
of Standard Oil's disposal, and that company's consequent
power to gain influence over other companies, the Committee
of Imperial Defence deemed that effective measures should be
taken to ensure that the Burmese oilfields should remain under
British control. That would necessitate keeping the directorate
and capital of Burmah Oil in British hands.[12]

The India Office therefore asked Burmah Oil to debar
foreigners from serving as directors or officers of the company
or from holding shares, much as the Cunard Steam Ship
Company had done since 1903. Eventually, at the beginning of
1907, the company offered to go as far as Cunard in accepting a
government appointee on the board with the right of veto over
any matters affecting the constitution of the company. In
return, it made its familiar requests for further concessions and
a promise of higher tariffs should there be another fall in prices.
Officials refused to consider a *quid pro quo*, and both proposals
foundered. No government director was appointed and the
Articles of Association were unchanged.

At the end of September 1905, Cargill, Wallace and Redwood met Miller at the Admiralty, and passed on the arresting news that Asiatic had, through an intermediary, offered them "a treaty of peace in the east".[13] Wallace and Deterding had been the subsequent negotiators, Deterding even taking time off from his very busy schedule to visit Glasgow and put some specific proposals to Cargill.

The initial terms were not attractive to Burmah Oil, and Cargill rejected them, even though the price war was depressing kerosene sales and profits alike. He did not disclose to Asiatic the negotiations over fuel oil, which had been conducted in total secrecy. After an interval, Deterding was back with new proposals on the basis of which Wallace and he swiftly reached an agreement over kerosene.

Asiatic was to withdraw entirely from Burma and Chittagong, and Burmah Oil from the Straits, China and Ceylon. In mainland India the company was conceded a prior market up to a maximum, but would have to limit its total annual production of kerosene. The difference between permitted production and actual sales in India and Burma would be sold to Asiatic at an agreed rate. That rate would cover little more than the cost of the crude and its refining, with only a small contribution to profit.

As important to the company was its insistence on fixing a maximum price for its Victoria kerosene. As later evidence shows, Asiatic objected very strongly to that maximum price policy as it tied that company's hands over prices, and demanded on more than one subsequent occasion that it should be relaxed.[14] Burmah Oil would not give way, and thereby greatly benefited the Indian consumer who between 1907 and 1910, when kerosene prices elsewhere were very high, was able to buy more cheaply than anywhere else in the world. In many speeches to his shareholders, Cargill reminded them of what the policy had achieved.

The 1905 agreement lasted in one form or another until it was replaced by a more comprehensive one in 1928. It has been portrayed as a resounding defeat for Burmah Oil, but not very convincingly.[15] As early as January 1903, the directors had accepted that, with increasing quantities of kerosene coming from the Syriam refinery, the Indian market could no longer comfortably absorb that output. Either an indefinite price war must follow, or some agreement with rivals must make provision for excess kerosene to be sold in markets well outside India. They now had that agreement as well as Asiatic's reluctant acceptance of the maximum price policy, which would also keep down the price of superior and inferior imported kerosene. Moreover, the initiative for the agreement had come from Asiatic; Burmah Oil had not had to sue for peace.

In one respect, the agreement *was* a reverse for the company. If it could have gone on sublimely disregarding threats from rivals, secure behind its tariff walls and official bans on foreigners, then it would have been impregnable. But its earning power was being drained away at a time when expenses were rising, particularly those for seeking new sources of oil to safeguard the fuel oil demands for the Admiralty. There were also as yet quite incalculable commitments in Persia, as explained in Chapter VII. It was hard to have to sell surplus kerosene to Asiatic at unremunerative prices, but that was a price that had to be paid for maintaining an orderly market in India itself.

Deterding's earlier demands since 1902, which would have severely restricted Burmah Oil's overall sales and production, had therefore been scaled down to what could be regarded as an acceptable compromise. An informed observer, Sir Thomas Holland of the Geological Survey of India, later summed up the kerosene agreement as follows: the company had signed with Asiatic "not a bond of friendship so much as a deed of separation" in which it "agreed to fight with gloves on in order to limit the damage to both parties."[16] Having fairly satisfactorily safeguarded his market in India, Cargill could turn with some assurance to his other outside negotiation, that over fuel oil.

In October 1905, Cargill informed the Admiralty of the agreement with Asiatic, assuring Miller that having secured more remunerative kerosene prices, he would be able to sell fuel oil that much more cheaply. Miller raised no objections, while Pretyman believed the agreement was "possibly the only way of preventing the Burmah Oil Company from being ultimately defeated". Wallace took a more positive line by stressing that, having reached peace with one opponent, the company would be "confident enough to sink in its oil works the necessary capital" for making naval fuel oil.

Nor did the then Secretary of State for India see the agreement as any barrier to granting the company further concessions in Burma. Only the viceroy struck a discordant note. Curzon was then in the final weeks of office, a disillusioned and embittered man after his enforced resignation. He refused to approve what he called "a combination of this nature" on the grounds that it harmed the Indian consumer's interests; he had of course overlooked the maximum price provisions. Yet he conceded that he had no powers to veto the agreement.

By then the Admiralty was becoming anxious that the fuel oil negotiations had still not been concluded. Balfour's cabinet was on the brink of resignation and its Liberal successors were unlikely to be so amenable. The contract was finally signed at the end of November 1905. In peace-time the Admiralty could draw up to 10,000 tons a year, and the company would build up a reserve stock of 20,000 tons against emergencies. In the event of hostilities, up to 100,000 tons would be produced: "a splendid war reserve", commented their Lordships approvingly.

Cargill had already planned to visit India and Burma and to travel out with Greenway who had just finished his leave in England. Adamson and Hamilton were to be in charge in Glasgow and Fleming in London, for Wallace was unfortunately once again in a nursing home with complications from his earlier operation. It looked as if Greenway might have to stay behind until Wallace was back at work, but he was able to start on the pre-arranged date.

On 25 November the pair left for India, the more easy in their minds now that the agreements had been safely concluded. Nine days later, the Conservative government resigned. It remained to be seen whether the new Liberal administration in Whitehall would wish to alter in any way the good relationships with ministers and officials which the company had succeeded in building up over the past two eventful years.

Notes

1 1937 AGM, *Glasgow Herald*, 5 June 1937.
2 The main sources of information in the public records are, for the Admiralty, PRO ADM 126/3807, and for the India Office, IOL L/E/ 7/498, 3044/03 and L/E/7/543, 2706/05. A very useful digest of official correspondence is in IOL L/E/7/608, 93/11 (filed under 438/08), "Ab-

stract of Principal Papers Relating to the Supply of Burma Fuel Oil to the Navy".

3 Command Paper Cd 1991, 1904: Second Report of Royal Commission on Coal Supplies II, minutes of evidence and appendices. Evidence of Miller, 18 March 1903, pp. 143–55, and of Redwood, 28 April 1903, pp. 190–213.

4 Under the maximum price policy, the company undertook not to sell inferior (Victoria brand) kerosene from its Calcutta installation at above Rs17½ (£1.17) a barrel, extra being charged for transport costs to places outside Calcutta. The price could of course fall below the maximum, depending on demand and supply.

5 IOL L/E/7/528, 43/05, 12 April 1904; cf. personal letters from C. K. Finlay about machinations of Asiatic and Standard Oil to Sir H. Barnes (Lieutenant-Governor, Burma) between 11 March and 17 August 1904, Bodleian Library MS Eng. Hist. e 260, Papers of Sir H. S. Barnes.

6 An obituary is in *Glasgow Herald*, 27 May 1904.

7 1905 AGM, *Glasgow Herald*, 27 April 1905.

8 1918 AGM, *Glasgow Herald*, 18 July 1918.

9 Sir A. Godley to J. W. Holderness, 1905, attached to papers of 1900 (see ch. IV, note 7 above), IOL L/E/7/441, 3436/00.

10 See R. Henriques, *Sir Robert Waley Cohen 1877–1952* (1966).

11 Ibid., p. 117. His Indian trip is described on pp. 108–24.

12 PRO ADM 116/3807: memo of CID, 10 August 1905.

13 IOL L/E/7/543, 2706/05: memo of interview between Cargill, Wallace, Redwood and Miller (Admiralty), 28 September 1905.

14 PRO ADM 116/1208: evidence of J. T. Cargill to Royal Commission on Fuel and Engines, 29 October 1912.

15 Notably by Gerretson, *History of Royal Dutch*. His account of events surrounding the kerosene agreement in vol. II, pp. 340–2 and Vol. III, pp. 50 and 211–13 are riddled with inaccuracies, e.g. the date of the agreement, that Burmah Oil (it was in fact Asiatic) had to withdraw from Chittagong and the names and specifications of the company's kerosene. He also makes the serious charge that Burmah Oil watered down its capital at the 1902 reconstruction by allowing ordinary shareholders nine shares for every one held, and in order to pay worthwhile dividends had deliberately refrained from building up financial reserves against future price wars, vol. II, p. 342: this allegation has been checked against the Dutch version, vol. I, p. 326. In fact, four, not nine, shares were allocated, and the board accepted the written-up value of its assets as a "moderate valuation"; that was not queried by Francis Palmer, then at the peak of his profession, nor by the auditors. In short, there is no evidence of an "imprudent" financial policy which was allegedly Burmah Oil's "one weak spot" and of which Deterding "successfully contrived to take advantage". Nor does Gerretson specifically mention the company's maximum price policy which proved such a thorn in Asiatic's side.

16 IOL L/E/7/642, 1023/09: minutes of evidence of Holland committee, Q. 282 Holland to Finlay, p. 126.

CHAPTER VII

Persia
1904–7[1]

When in November 1905, John Cargill embarked on his trip to the east, there was a third agreement he had concluded that was also destined to set the company on a new tack. Signed the previous May, that agreement concerned Persia, and the present chapter will relate the origins of Burmah Oil's interests in that country and events there up to 1907.

As in Burma, travellers had for centuries been sending back reports of Persian oil traces; yet scientific investigation alone could reveal whether oil existed in commercial quantities. In one important respect, conditions differed in the two countries. For centuries the Burmese had been raising oil from shallow wells, whereas the Persians merely baled out the seepages. Thus prospectors had to start virtually from scratch.

Among the first geologists to inspect oil-bearing lands in Persia was one of Boverton Redwood's men, H. T. Burls. Redwood sent him out in 1901 to investigate the two areas that seemed most hopeful: Chiah Sourkh in the north-west, and near Ahwaz in the south-west. This mission was on behalf of W. K. D'Arcy, a Devon man who had made a fortune out of gold in Australia, had settled again in England and was currently interested in a possible concession from the Persian government. Burls's findings, although based wholly on surface indications, were positive enough for D'Arcy to open negotiations in Teheran through various intermediaries. In May 1901 he secured a concession for sixty years, the shah authorizing the agreement personally. The concession was unique in the vast extent of territory involved, covering 500,000 square miles and embracing the whole of Persia apart from the five northern provinces.

As D'Arcy's consultant, Redwood chose an engineer to take charge of the fields operations: George B. Reynolds, then in his late forties and a former employee of the Indian public works department, who had gained drilling experience in the Royal Dutch fields at Sumatra. Chiah Sourkh was the first location he was instructed to test, in Persia but not far from Baghdad in what was then Turkish Mesopotamia.

Yet communications were far more difficult there than in the Burmese fields. Every single piece of machinery and equipment had to be shipped to Basra on the Persian Gulf, and 300 miles up the Tigris to Baghdad. It was then manhandled over the mountains to the oil site. Thus although Reynolds reached Persia in September 1901, he was not able to spud in the first well until November 1902. Even so, only a man of his physical endurance and sheer force of character could have achieved what he did in that space of time. Meanwhile, the cost of the venture was mounting steadily, while the hoped-for returns were being pushed further away into the future.

For D'Arcy those delays were a constant headache. He was no simpleton; he had become rich from gold and shrewdly backed his hunch over oil in Persia. Yet he enjoyed lavish spending on the material luxuries that abounded in Edwardian Britain. Not that his life-style was in the raffish jet-set manner of Deterding: he fitted snugly into the society of his day, being a public-school man (Westminster, no less) and pursuing a highly acceptable gentleman's mix of sporting and cultural activities. He gave shooting parties on his Norfolk estate and owned the only private box at the Epsom racecourse, while his actress wife, Nina Boucicault, both graced the London stage and held musical soirées at their London home in Grosvenor Square, to be entertained by such performers as Caruso and Melba.

He had a reassuringly bluff exterior and a slow-moving mind, again in contrast with Deterding, and therefore struck an answering chord in the hearts of ministers of the crown and top Whitehall officials whom he met. No one was likely to accuse him of being too clever by half. This temperamental affinity was to be an asset beyond price to him over the next few years, for the early months of 1903 found him heavily overdrawn on his

London bank as cash drained out on the Persian venture. Luckily for him, the concession imposed no other financial commitments until such time as oil was struck. He merely had to set up an operating company and pay £20,000 to the Persian government within two years of May 1901.

That operating company, the First Exploitation Company Ltd, was registered just on time in May 1903, the rather odd name being lifted from the relevant clause in the concession agreement. D'Arcy became chairman, with two other directors and Redwood as technical adviser. The issued capital was 350,000 £1 shares of which D'Arcy received the bulk in exchange for the concession rights, but 20,000 were given to the Persian government over and above the cash payment mentioned above; some shares went to officials in Teheran.[2] None were offered to outsiders, as there was no prospect of paying dividends for the foreseeable future. The company's formation did nothing to staunch the drain from D'Arcy's pocket.

Finding his money troubles too much for him, he offered to sell the whole or part of the concession to various financial groups in London. However, oil failures were too common in too many parts of the world for potential backers to show much

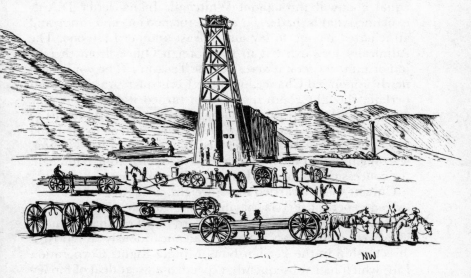

Iron pipe transporters – jims – at D'Arcy's well, Chiah Sourkh, c.1905.

enthusiasm until positive results had been achieved. In October or November 1903, when the two wells at Chiah Sourkh were down to nearly 200 and 250 feet respectively, he spoke to E. G. Pretyman, the parliamentary secretary at the Admiralty. (See Chapter VI.) He asked whether the British government would be prepared to make him a loan in return for an undertaking to supply naval fuel oil in due course.

The idea clearly came from Redwood who, as a member of the Admiralty's Oil Fuel Committee which Pretyman chaired, masterminded its energetic quest for reliable sources of oil; indeed, members then awaited a reply from Burmah Oil to their initial enquiries of the previous July. Although Persia was a fully independent country, its southern part, like the Persian Gulf, was effectively under British influence. Pretyman therefore advised D'Arcy to write to the Admiralty, outlining the course of events to date and specifying how much might be needed to place the concession on a strong financial footing. Selborne, the First Lord, and he would then set about persuading the Treasury to authorize a loan secured on the concession. In return, the Admiralty would require a fuel oil contract.

That startlingly simple advice had one fatal flaw: it assumed equal goodwill throughout Whitehall. In his letter D'Arcy explained that he had so far spent £160,000 on exploration and anticipated having to lay out at least another £120,000. The Admiralty forwarded it to the Foreign Office; from there a junior minister took it across to the Treasury. The young and newly appointed Chancellor of the Exchequer, Austen Chamberlain, thought about it and then turned it down, since he could see no chance of persuading the House of Commons to vote such money. The junior minister then went on to the India Office, which immediately cabled the viceroy, Lord Curzon, warning him that if no loan were forthcoming from Whitehall, the concession might well fall under Russian control.[3]

Curzon's views carried much weight, as he was an acknowledged expert on Persia, having travelled extensively there in 1889–90 and then written a standard work on that country. However, a few years later he had incautiously accepted a directorship in the Persian Bank Mining Rights Corporation Ltd, which had subsequently expended a great deal of money on rather haphazard drilling for oil before going into liquidation, and that débâcle had bitten very deep. When in 1901 the India Office had informed him of the D'Arcy concession, his damping reply had been that he saw no chance of Persian industrial regeneration making a new start in D'Arcy's hands.

Now, in December 1903, his reply to the Foreign Office was that the Persian oil deposits were unlikely ever to be worked at a profit. With characteristic ambiguity of mind he added that he could see no risk in the concession being sold to Russia, but for the prospect that it would be worked in conjunction with the wells at Baku on the Caspian sea, and therefore immeasurably strengthen Russian interests in the British sphere of influence.

That was precisely the point at issue. He had touched on one of the most recurrent themes of Britain's foreign policy outside Europe: the security of her communications with India. Persia, then close to political and financial collapse – like Upper Burma in 1885 – seemed ripe for absorption by a great power, and to the north Russia had already made preparatory moves. The latter's ultimate goal was to secure warm-water ports on the Persian Gulf; since 1900 it had taken some preliminary steps, such as setting up consulates and seeking to establish a coal-bunkering service in those ports.

The Foreign Office, much more alarmed than the viceroy, despatched Sir Arthur Hardinge, British minister at Teheran, to inspect the oilfields at Chiah Sourkh and assess their possibilities. Arriving as he did when some important oil shows had already occurred, he reported back in January. Cordially disagreeing with Curzon's "pessimism", he urged at least a British majority interest in the D'Arcy concession, as well as strict safeguards against allowing it to pass into foreign hands. One of the Russian banks in Teheran, he pointed out, would be only too glad to buy the concession regardless of its commercial value, as a means of extending Russian influence to the whole of Persia and on into neighbouring Mesopotamia. The shah might then authorise the Russians to build a pipeline down to the Persian Gulf.

That was a realistic assessment, for as recently as 1902 the Russian Minister of Finance, Witte, had striven hard to secure a pipeline concession across Persia in return for a tempting loan to the bankrupt shah. Yet he had failed to win for his country a foothold on the Persian Gulf; a failure later described by a Soviet historian as "one of the first serious defeats of Witte's

economic policy in Persia" in the period preceding the Russo–
Japanese war of 1904. "Only a few years were needed," the
historian continued, "to enable the Russian government fully
to appreciate the significance of this defeat."[4]

But for Britain the threat remained. In May 1903 Lans-
downe as Foreign Secretary had made a historic declaration to
the effect that his government would regard any naval base or
fortified port established by another power in the Persian Gulf
as "a very grave menace to British interests", one that his
government would stoutly resist. Yet almost a year later he had
no serious ideas for countering an equally grave economic
menace. Were Persia to fall to the Russians, then Mesopotamia
might be the next to go, and after that, then India. . . .

Early in 1904, the oil shows at Chiah Sourkh turned into
flowing oil. D'Arcy, never more hungry for glad tidings, was
moved to comment ecstatically, "Glorious news from Persia
and . . . the greatest relief to me." His euphoria was soon
dashed when the abundant flow declined and, even more
worryingly, brine was found to be mixed in with the oil. Redwood had to explain that the wells were most unlikely in the
long run to repay their cost. Even if they did, there remained the
very difficult question of transporting the oil in that moun-
tainous region 400 miles from the sea. In any case, the geologists were coming round to the view that oil in quantity could be
discovered only in south-western Persia, where far fewer pipeline and communications problems existed than in the north.
He therefore advised D'Arcy to have the wells at Chiah Sourkh
shut down.

The financially hard-pressed D'Arcy at once had to seek
outside help. The Paris Rothschilds, with their one-third stake
in Asiatic and interests in Russian fields, were heavily involved
in oil and currently supplied more kerosene to the Indian
subcontinent than either Burmah Oil or the Americans did.
However, when D'Arcy met one of the Rothschilds in France,
they could not hit on a basis of agreement. It was therefore some
comfort to receive a letter from Pretyman requesting him,
before he finally decided to sell out to foreigners, to give the
Admiralty a chance to try to form a consortium of British firms.
Instead he turned to a number of other parties, including an
unnamed American group and the very wealthy financier, Sir

Ernest Cassel. None of them expressed any real interest and, by the summer, D'Arcy was at the end of his tether.

Yet his needs coincided exactly with those of the Foreign Office, anxious about the route to India, and of the Admiralty, seeking reliable fuel oil supplies: they required a British figure of affluence and patriotism, willing to shoulder the considerable risks involved in proving the existence of oil in Persia. Ideally, a prominent imperialist was needed to head a syndicate of patriots: during the summer, therefore, the First Lord made contact with one of the wealthiest self-made men in Britain, Lord Strathcona. Then 84 years old, he was a well-known imperial figure, although not of the first rank. He had amassed his fortune in Canada and was the dominion's High Commissioner in London. Although he gave a million or more away in charitable benefactions of one kind or another, he was not one to squander money on risky commercial schemes: he was therefore unlikely to prove an especially soft touch. However, Redwood's sure hand can be detected here, persuading Selborne to instruct Pretyman to sound out Strathcona.

The aged imperialist's initial reaction was cool but not unencouraging. "I rather smiled at the idea of being asked to take it [the concession] up in the first instance," Strathcona later recalled: ever since his retirement at the age of 75, he had refused to become involved in any new commercial project.[5] However, he continued, "Lord Selborne was very insistent indeed, so I consented to do it." He gave way only because he looked on the proposition not from the commercial "but really from an imperial point of view". Such indeed had been the devout hope of ministers. Pretyman himself remembered that Strathcona asked only two questions. Was it in the interests of the Royal Navy that the enterprise should go forward and that he should take part? Most certainly, replied Pretyman. Then would it meet the case if he were to invest £50,000 in it? Pretyman said yes, hastening to reassure this multi-millionaire that it was not so much financial assistance as his publicly expressed support that was needed. So Strathcona's role was a passive one; he was not required to go out and form a syndicate of his own.

Where, then, was the syndicate to be found? Someone took the initiative in approaching Cargill as chairman of Burmah Oil during July or August 1904, and it would be instructive to know who that someone was. Clearly, it was not Admiral Sir John Fisher, although a year earlier, in July 1903 – two months after the formation of the First Exploitation Company – he had met D'Arcy at the Bohemian spa of Marienbad. When D'Arcy revealed that he was looking for someone to "manage" the concession, Fisher was interested enough to ask to see all the relevant maps and papers, but in the end failed to grasp the political and strategic significance of Persian oil. The admiral's invitation, in the spring of 1904, to D'Arcy for a weekend stay at the commander-in-chief's residence at Portsmouth seems to have been purely a social one, and in the "scheme" he wrote on taking office as First Sea Lord that October, he dismissed the adoption of oil propulsion as merely one of "many, many projects" and quoted approvingly Selborne's views as First Lord that oil did not exist in the world in sufficient quantities and must, therefore, be used only as an adjunct.[6]

Another possible intermediary was Pretyman, who by that time knew Cargill and Wallace quite well through the fuel oil negotiations. Yet once again the obvious go-between was Redwood, as one of Burmah Oil's most prominent top managers of the next generation, R. I. Watson, later surmised.[7] As a consultant to both Burmah Oil and the First Exploitation Company and a member of the Oil Fuel Committee, Redwood was a unique link in this momentous chain of negotiations.

Whoever may have brought them together, Cargill and Wallace met D'Arcy for the first time on 10 August 1904, Pretyman acting as host. D'Arcy was overjoyed to find, after his disheartening parleys with other magnates, that the directors seemed "to want to do business". They initially concerned themselves not with large ideas about imperial interests or command of the seas, but with the exact political status of Persia and whether it could be considered as being under British protection or not: the earlier frustrations over oil supplies from the weak but politically independent state of Upper Burma were less than two decades away. Pretyman referred them to the Foreign Office, where a few days later a senior official was able to go some way towards reassuring them. Later, a written enquiry elicited from the Foreign Office a favourable opinion about another vital point, the legal validity of the concession.

D'Arcy rushed in with mention of some fairly hefty sums of

money. In his relief at finding potential backers who were both in earnest and entirely above-board, he threw out the figures of £500,000 as the likely capital and the £150,000 to repay his own expenses to date in exchange for half the First Exploitation Company shares. "This is, of course, quite satisfactory to me," he added; as it well might have been. Again following precedents in Burma, before even discussing finance Cargill and Wallace asked to see all the geological reports. Those were quickly made available, including the crucial one of 1901 by Burls with Redwood's annotations.

At first Cargill was extremely reluctant to become involved. Bearing in mind the load of work that still needed to be done in Persia, he felt that Burmah Oil already had more than enough problems to exercise him. Wallace was made of sterner stuff. Steeped as he was in Indian commercial affairs, he well appreciated the danger to the whole petroleum trade in India should Britain lose the Persian concession. Over and above the strategic menace from Russia, any attempt by foreigners to flood India with Persian kerosene would perhaps irretrievably damage Burmah Oil's future.

For their separate reasons, Cargill and Wallace did not rush over D'Arcy's proposal. The penalties of losing the concession were real enough; yet if they decided to take it over, the fact was that Burmah Oil's finances were none too strong at that time, after more than a year of the kerosene price war. D'Arcy, not appreciating the deliberate way in which their minds worked, was surprised that they did not arrange early meetings with Reynolds – then on leave in England – and other experts to discuss points of detail. He had to wait until the end of November before they decided in principle to go ahead.

A preliminary accord was initialled the following January, after Cargill had for the first time informed his board of the negotiations. He consented to bear the expense of proving the oil in southern Persia and to find up to £1,000,000 in capital to develop the concession into a company. Already engaged as they were in one major negotiation, with the Admiralty over fuel oil, Cargill concentrated on that and left Wallace to treat with D'Arcy; his private secretary attributed the successful outcome of the latter discussions to Wallace's unfailing resourcefulness and tact.

As soon as that agreement had been concluded, Reynolds

was brought in to discuss with Redwood how much would be needed for prospecting. Reynolds estimated that £40,000 would be more than enough for tests in the south. That was rounded up to £50,000 and the company added a further £20,000 as a "safeguard". The agreement would lapse should no oil be found; D'Arcy would then have to repay his £25,000. If the venture were successful, a £2,000,000 company would be formed, Burmah Oil having to underwrite the issue of £800,000 worth of debentures.

The official agreement with D'Arcy was signed on 20 May 1905. That set up a company to deal with Persia, the Glasgow-based Concessions Syndicate Ltd, effectively an offshoot of Burmah Oil's plant and stores department there. The three directors, Cargill, Adamson and Hamilton, were intended to act as a steering group. Here was yet another interlocking unit in the Glasgow office, with capital of £100,000, of which Burmah Oil subscribed £95,000. Strathcona acquired the rest, his holding being purely symbolic since £5,000 was neither here nor there to Burmah Oil. The Concessions Syndicate undertook to spend up to £70,000 over the coming three years on prospecting work, and to pay D'Arcy £25,000 in cash.

The First Exploitation Company became a subsidiary of the Concessions Syndicate, with D'Arcy transferring to the latter 300,000 of his 350,000 shares. He remained as chairman, but his two fellow-directors were removed to make way for Cargill and Wallace. While the Concessions Syndicate in Glasgow would be responsible for plant, stores and finance, its subsidiary, being in London, would handle all diplomatic and political questions. That roughly mirrored the division of responsibilities between Burmah Oil's two offices.

The previous month Cargill and Wallace, together with Redwood, had completed their very detailed survey of the geological reports. They agreed with the geologists that future exploration should be in the south. That would be simultaneously at Shardin, on the foothills of the Zagros mountains some 85 miles north of the Persian Gulf, and at Masjid-i-Sulaiman, otherwise known as Maidan-i-Naftun, an inhospitable place about 55 miles from Shardin and 120 miles from the Gulf. Reynolds would be in charge, and the directors briefed him as thoroughly as they were accustomed to brief their other experts on the spot. There is no evidence, as has been suggested, that they cold-shouldered him at the time, or that any offhandedness of theirs led to subsequent antagonism.[8] The probability is rather that this proud and independent man may already have guessed the degree of interference to come.

Early in April, shortly before departing for Persia, Reynolds attended a final conference to fix the quantities of plant and equipment required at both sites. Anxious as ever to leave nothing to chance, Cargill and Hamilton made him set out those requirements in writing. In a letter of reply, Cargill gave precise instructions on matters ranging from the methods of surveying to be followed to the grounds on which he could dismiss drillers. Cargill granted him permission to write privately on any sensitive or confidential issue.

Reynolds's activities day by day, during his three-year tour of duty from 1905 to 1908, can be followed from the letters and cables he sent regularly to the Concessions Syndicate.[9] In July 1905 he reached the already abandoned site of Chiah Sourkh, plugged the wells, and had the two drilling engines locked away in a workshop. The remaining machinery, carts and camp equipment, all 40 tons of it, he took south with him via Baghdad, down the Tigris to Basra and across to the port of Mohammerah (later Khorramshahr). A native boat loaded with this gear was towed by a launch belonging to a local British mercantile firm, F. C. Strick & Co. Meanwhile, a party of twenty men were "marching" the 125 mules and horses down, to save the heavy cost of transport.

For the time being Reynolds operated from Mohammerah, while permission was sought from the local tribesmen, the Bakhtiaris, to drill, and also an agreement not to pilfer the equipment; for the shah's writ did not run there. As negotiator, the Foreign Office provided J. R. Preece, consul-general at Isfahan which was in Bakhtiari country. Once Preece reached an agreement, Reynolds was at last ready to move. He deposited at Shardin an Arab surveyor to chart the various geological formations below the surface that might indicate the most hopeful sites. Going on to Ahwaz, the town nearest to the two sites, he called up the plant and the remaining horses and mules from Mohammerah.

That December, with commendable foresight, he wrote personally to Cargill recommending a site for the future refinery. He had chosen Abadan, a long and narrow island of mud flats and palm groves on the Shatt-el-Arab, the delta of the Tigris and Euphrates, some 30 miles from the Persian Gulf. He observantly noted that the water on coastal side of the island was 10° cooler than on the Shatt-el-Arab side, which could only benefit the condensers while refining.

Not until April 1906 was he able to send home a tracing of the Sharin survey plan, to allow a choice to be made of the initial sites for drilling. He received instructions in June, just about a year after his return. In August, No. 1 well was started, the second following two months later. Drilling proved to be painfully slow, the unsatisfactory fuel – brushwood in an area notoriously short of timber – making it almost impossible to keep steam up. At first drilling averaged no more than 50 feet a month. As no oil was discovered near the surface, they would have to wait until the drillers had gone through the next layer of thick gypsum beds and reached possibly oil-bearing rocks.

For those at home, that delay seemed inordinate and was extremely trying. Only the previous September D'Arcy had over-optimistically forecast that oil would have been dis-covered before the advent of the hot season in May slowed operations up. That month D'Arcy invited Wallace and his daughter to watch the Derby from his box near the winning post. No doubt the luxurious food and drink upset Wallace's delicate liver and helped to bring on no less than four "painful attacks" in the next few weeks. Despite further invitations, he did not attend the Epsom races again.

Nor was it a coincidence that relations were steadily de-teriorating between Reynolds and Glasgow. At the outset he had been on his best behaviour, trusting in his principals' long-standing experience of technical matters and of conditions in the tropics: qualities D'Arcy had rather lacked. That Reynolds's main correspondent in the Glasgow office was James Hamilton turned out to be unfortunate. As a director of the Concessions Syndicate – although not yet of Burmah Oil – he knew all about plant and stores questions. But he never really discarded the outlook of a chief clerk and (uniquely among his colleagues) had no experience of the east: a great deal of what was happening in Persia was therefore outside his understanding altogether. To be fair, Glasgow had never before dealt directly with a man in the field. In Burma technical matters were normally filtered through Finlay Fleming & Co. whose senior men perfectly understood the directors' thinking. Yet there was no excuse for Hamilton to treat Reynolds as a functionary or to withhold from him the kind of courtesy his responsible position entitled him to.

The first major fracas was about the steel casings used to line the wells and prevent caving in. The various lengths were screwed together as they were sunk in the holes, and the threads were usually protected at each end by steel caps. Reynolds, having already stressed to Glasgow the absolute need for adequate protection for the threads, was dismayed when in December a consignment arrived with only wooden plugs at each end, and the threads so badly mangled that they would have to be rethreaded on the spot. Drilling was therefore held up until rethreading tools could be sent out from England.

Instead of making sure that it would not happen again, Hamilton tried to bluster it out, refusing to accept that so many of the lengths had been damaged, and later suggesting that "extraordinarily bad handling" in the course of unloading in Persia had been responsible. In fact, Reynolds had personally supervised their unloading. Equally unfortunate were Hamilton's reactions when Reynolds had to sack a Canadian driller, C. H. Locke, for striking a Persian while drunk. Hard liquor was forbidden for the drillers (but not for senior staff) and a local mechanic had smuggled some in.

On his way home through London, the directors questioned Locke about his dismissal and, less ethically, about the damage to the casing lengths. He predictably claimed that he had been wrongly dismissed and that the casings must have been damaged after their arrival in Persia. Reynolds's retort was to the point. "The statements of this man to you appear to denote a rather careless handling of the truth" and, later, "that the writer [James Hamilton] should give credence to the word of a man who was discharged and who denies it is, if I be permitted to say so, astonishing!"

To crown it all, there was the distressing incident of the surveyor. Reynolds asked for an assistant engineer, preferably a trained Indian, to lay down the road to the other earmarked site, Masjid-i-Sulaiman. Those at home asked the government of India to find one; only after protracted correspondence was one despatched. He arrived at Karachi, went sick and was never heard of again. Eventually Hamilton had to instruct Rangoon to send one over and a British technician was chosen. W. G. Parsons had the distinction of being the first Burmah Oil employee to serve in Persia: the sole distinction in an otherwise unrelieved chronicle of failure.

On his arrival Parsons created a singularly poor first impression by confessing to Reynolds that he was no surveyor but rather good at lifting heavy weights. He went on to make a badly botched job of the road, giving it such gradients and sharp turns that only mules, not carts, could use it. No wonder he started to drink heavily and had to be sent back to base with an indisposition. Then a cable arrived from Glasgow, saying that he was urgently needed in Burma, and enquiring when the road would be completed. On the strength of a mutilated word in the cable, Reynolds chose to take it as an order to return him forthwith. By February Parsons was out of Persia.

A month later the inevitable comeback arrived from Glasgow: "We note you will since have learnt that the haste was not so necessary as you read into our telegram." Reynolds did not mince words in his reply, spelling out the man's incompetence and also his boasts about having been sent to report on conditions in the country. Those boasts were false, but Parsons does seem to have been told to write privately to Rangoon, so that his findings could be forwarded in confidence to Glasgow. Yet in his cups he had revealed that he was there as a punishment for having killed a Burmese, who had shot one of his men first.

He had not yet finished plaguing Reynolds for, when out of Persia, he claimed that a driller had died not of pneumonia but from want of food. Reynolds angrily deplored Glasgow's "sympathetic acceptance of statements from irresponsible parties". His fury was so intense that his final sally was unusually feeble: "Have you not afforded me the consolation of religion, in that you sent me Parsons? Faith, a surveyor had been better!"

A perennially sensitive issue was that of food. Reynolds once likened himself to the skipper of a windjammer, constantly plagued by complaints about the rations; those who really understood life under sail – Joseph Conrad, for instance – knew that a hard but scrupulously fair taskmaster was likely to get more out of his men than one who drove himself to the limit and constantly took it out on those striving to match his impossibly high standards. In some ways the monotony, physical hardships and deprivation in the Persian fields were comparable with those endured by seafaring men, but the drillers and others were too often bawled out by this large and terrifying man to have any abiding loyalty to him.

Reynolds was no monster. He had a wife, to whom he wrote regularly, a pet dog and his pipe; he was considerate enough to provide cider, library books, gardening seeds and the game of bowls for men deprived for years on end of the most basic comforts of civilization. He was also particular about ordering from Britain a wide variety of tinned and other foodstuffs and searching for whatever fresh foods were available locally. Yet an elderly driller with the Conan Doyle-ish name of Tinswood Slack, when invalided home and interviewed at the London office in December 1906, claimed that the rations were very meagre and of poor quality, with potatoes and vegetables often lacking. To show the inferior quality of Persian bread, he laid on a director's desk a piece he had brought back with him.

When challenged with these assertions, Reynolds remarked with justice that Slack had never made a complaint in Persia, and that any bread might look rather peculiar after a month's travel. Moreover, the drillers were incapable of managing for themselves, despite having competent cooks. "And now," he concluded in a not entirely unfriendly way, "thanking you for your homily on the treatment of men, let me assure you that the men have very little to growl at, and if only they would look after themselves a little and their interests instead of sitting down and growling, they could do better."

Hamilton did not have the sense to drop the matter. "We fear there is more in it than appears on the surface," he replied. Reynolds pressed him hard for the precise evidence, but only received the weak response that "apparently the food question was not quite what it was stated to be by the men who came home," to which he retorted, "You evade giving me a direct answer to my queries." All this slanging proves beyond doubt the dangerously thin top management in Burmah Oil during 1906–7. Wallace was preoccupied with his day-to-day contacts with Cohen (see Chapter VIII), in addition to dealing with First Exploitation Company matters and, in between times, running his own agency of R. G. Shaw & Co. Mercifully, his

health stood up to that burden. Cargill and Adamson were fully
occupied with managing the various Cargill interests in Glasgow, and that left Hamilton striving to cope with Reynolds in
between all too frequent trips to the inadequately manned
London office.

One tiff at least brought Reynolds an agreeable outcome.
That was about the camp doctor, Dr Rustom Desai, a touchy
and sometimes neurotic character who was a thorn in
Reynolds's side. They managed to endure each other until, in
March 1906, Desai demanded one of the very limited string of
horses for his own use. He then refused to treat a wounded
tribesman on the ground of lacking some items of essential
equipment, an excuse dismissed by Reynolds as frivolous. In
the subsequent row Desai imprudently declared he would be
only too glad to get out of Persia. Reynolds took him at his word
and at once cabled Glasgow for a replacement.

Hamilton argued that there must be something deeper than
merely what he called "the horse question", which seemed far
too trivial to be referred home; besides, he added, trouble was
only to be expected from the tactless way Reynolds had handled
the affair. Reynolds's return shot was straight between the eyes.
"You really amuse me by instructing me how to run a contu-
macious Parsee [Desai] and an alcoholic driller [Locke], both
suffering from swelled heads." It riled him even more to hear
later that Hamilton had bought off both of them, Locke with an
ex gratia payment of £50 and Desai with £120.

Not until June 1907 did a replacement come: Dr Morris
Young, who had lately turned down the post of house surgeon
at the Glasgow eye hospital. Educated in the cosmopolitan
Syrian town of Aleppo – his original surname having been
Yudlevitz – he knew quite enough of human nature not to be
cowed by Reynolds's surly remark on his arrival: "You've been
a devil of a long time getting here!" The tactful and highly
gifted Young became a close personal friend of Reynolds, and
certainly contributed to the easier relations with home that
existed in the critical twelve months that followed.

Notes

1 For the present chapter and chs IX and XII on Persia, I have relied in part
on a privately printed (*c.*1937) but unpublished and untitled narrative of

events relating to Anglo–Persian and its origins in 1914. It was attributed to Hilary St George Saunders by H. Longhurst, *Adventure in Oil: The History of British Petroleum* (1959), p. 10, but the author is now believed to have been Dr Laurence Lockhart. Longhurst's book, although about to be superseded by Dr Ferrier's authoritative history of BP, provides a very readable and evocative sketch of that great company's past. Anglo–Persian's abridged prospectus, in *The Times*, 17 April 1909, is also an important source.

2 The exact number of shares given for services rendered is not known, but in July 1908 the issued capital was £397,000, of which £350,000 were D'Arcy's original shares.

3 The story is told in IOL L/PS/3/403, 3332/03 and 3368/03, December 1903.

4 B. V. Anan'ich, "Rossiia i kontsessia d'Arsi" (Russia and the D'Arcy Concession), *Istoricheskie Zapiski*, 66, (1960), p. 289, quoted in F. Kazemzadeh, *Russia and Britain in Persia 1864–1914: A Study in Imperialism* (1968), p. 385.

5 Admiralty Committee on the Use of Fuel Oil in Navy I. Departmental and Other Reports and Minutes of Evidence, December 1911–February 1912 and Interim Report of Committee, 19 January 1912 (25 April 1912) in Naval Historical Library. Evidence of Strathcona 29 December 1911.

6 A. J. Marder (ed.), *Fear God and Dread Nought: Correspondence of Admiral of the Fleet Lord Fisher of Kilverstone*, I, (1956), p. 275. cf. *Fisher Papers* (ed. P. K. Kemp, Navy Records Society I, 1960), pp. 80–1, "Scheme" 21 October 1904.

7 In a pencil comment scrawled across p. 49 of the unpublished history of Anglo–Persian: "Boverton Redwood was undoubtedly the connecting link."

8 This is the view of R. W. Ferrier, "Makers of BP No. 2, G. B. Reynolds of MIS", *BP Shield International*, May 1972, p. 14.

9 The four volumes of the original "Letters G. B. Reynolds to Concessions Syndicate Ltd." are in the Burmah Oil archives, running from 2 July 1905 to 2 June 1909.

CHAPTER VIII

A Period of Transition (1)
1905–8

The events of November 1905, involving the Admiralty fuel oil contract and the kerosene agreement with Asiatic, radically altered the framework within which the Burmah Oil Company operated. Never again would it have an almost total freedom to act as it chose in the market, with only relatively minor commitments such as its jute batching oil contracts. Instead, it was to be constrained by permanent obligations which forced it to pay more systematic attention to the securing of crude oil supplies and economies in all its activities.

Much of 1906 was taken up with two measures of reorganization: overhauling the financial arrangements with Finlay Fleming & Co. in Rangoon,[1] and doing something about the London office. The former arose out of the pressing need to scale down the very high payments to the Rangoon agency, which received a 5 per cent commission on direct sales and 2½ per cent on sales through sub-agents such as Shaw Wallace & Co. in Calcutta. Soaring outputs of oil products had pushed up these payments from £23,000 in 1900 to £49,000 in 1905, equivalent to over a quarter of net profits. As John Cargill and Adamson had conflicting interests here, being partners in Finlay Fleming & Co., the tidy-minded Wallace took the initiative and persuaded Cargill, while on his eastern tour of 1905–6, to renegotiate the terms so as to reduce the burden on Burmah Oil. The solution Cargill brought back was to pay a fixed fee of £15,000 a year and a 2½ per cent commission on gross profits.

Because the change entailed amending the Articles of Asso-

9. Kirkman Finlay (1847–1903), first managing director.

10. Charles William Wallace (1855–1916), director 1902–1915. From a portrait by Orpen.
(Reproduced by kind permission of Brigadier R. M. Carr)

11. Yenangyaung field, early twentieth century. Burmese well owners (Twinzas) delivering oil at depot.

12. Local transport of oil drums at Yenangyaung.

13. Assam Oil Company's refinery under construction at Digboi, 1905.

14. Thatched oil derrick (Digboi) 1905. Protection against the elements for all-year-round operations.

15. Foreshore of the Irrawaddy at Nyaunghla, Yenangyaung with town in background. The larger boats were used for long distance travelling on the river.

16. Local guards, Syriam refinery, pre-1914.

ciation, the obvious move would have been to call an extraordinary general meeting directly after the annual meeting in May. The directors failed to ratify the new agreement in time, and that gave disgruntled shareholders full scope to air their grievances in the correspondence columns of the *Glasgow Herald*.[2] They alleged extravagant management in the east and insufficient information about agency activities. Moreover, they claimed, rates of dividend were too low, given the tariff advantage enjoyed in India. Therefore what was needed was new blood on the board in order to deal with these problems and keep abreast of the times.

As there happened to be two vacancies on the board, a battle ensued over how they should be filled. When the company nominated James Hamilton and a retired manager of the Bank of Scotland, Robert Gourlay, a protest meeting of shareholders objected to Gourlay as knowing nothing about oil and to Hamilton for being the company's stooge. Instead, the meeting proposed two Scottish shale-oil magnates, one being William Fraser of the Pumpherston Oil Company, whose son and grandson were to achieve great prominence in the British oil industry (and incidentally become Burmah Oil directors) as the first and second Lords Strathalmond. A rival shareholders' group, however, ridiculed these names because the shale-oil companies were direct competitors in products such as wax and jute batching oil.

To press their attack home, eighty dissident shareholders met and passed a resolution to block both of the company's nominations and also Cargill's new proposals for the agency commission, which they held to be still too generous. As one speaker put it, "Scottish shareholders have never been used to allowing directors an absolutely free hand." He also deplored the close links that bound the company to its Rangoon agency, whereas there ought to be a strictly arm's-length relationship.

The extraordinary general meeting in July proved very contentious, with Cargill strenuously denying that shareholders had any right to dictate names to directors. However, no other nominees were put forward and his candidates were duly approved. Two months later a further meeting was held to agree the new agency commission; a move to reduce the fixed fee to £5,000 was defeated and the matter went through. It was also agreed that after five years, Finlay Fleming & Co. could at any time be removed from the agency by resolution of a general meeting, without having to prove incompetence or wilful default. That amendment was destined to be used in the late 1920s.

Another of the company's functions to be reorganized in 1906 was the London office. Above all, it now required a top manager to maintain day-to-day contact with Asiatic on matters arising from the kerosene agreement, as well as extra clerks to keep records of how it was working out. Wallace was the initial contact man, but chose to operate from his own firm of R. G. Shaw & Co. rather than move permanently into Burmah Oil's London office. He detested that office, which was so poky that it had no board-room; during all the negotiations he had conducted since Kirkman Finlay's death in 1903, he had therefore held meetings in his own premises.[3] Partly also, having declined the managing directorship, he wanted to avoid as many of the routine chores as possible. In fact, as he was already preparing for semi-retirement, that year he brought H. S. Ashton, senior partner of Shaw Wallace & Co., home from Calcutta to assist him in R. G. Shaw & Co. Greenway took over in Calcutta as senior partner.

Then two fortuitous events brought about changes, although of a halfway rather than a radical kind. For internal reasons, the London office's oil activities had to be separated from Milne & Co.'s non-oil responsibilities after Hamilton had been promoted to "manager" in Glasgow and the share registrar, F. G. Holdsworth, became company secretary. Wallace therefore was asked to lease for the company new and larger premises – with a board-room – at 93 Bishopsgate. The other event arose from the company's maturing relationship with the Assam Oil Company, registered in 1899 after Burmah Oil had refused to acquire the assets (see Chapter IV). As will be shown in Chapter XVII, Burmah Oil did not acquire a controlling interest until 1921; but in 1906 Assam Oil happened to make an agreement with Asiatic over its surplus kerosene on the same terms as with Burmah Oil. Since the arrangements were interlocking, the two companies decided to have representatives on each other's boards.

Ashton was Burmah Oil's nominee to Assam Oil's board. A Lancashire man and a great footballer, he was the first Englishman to attain a high position in Burmah Oil. However, he and the first Assam Oil nominee, Sir Thomas Bowring, were not appointed to the Glasgow board but to a specially constituted London board. The latter was a purely formal one with no distinct functions of its own. It met irregularly and had such scanty business that after a time it ceased to meet at all, and the handful of subsequent board appointments was made primarily to confer some status on those concerned. When Bowring resigned in 1912, he was not replaced, but the London board limped on until 1928. Like Wallace, Ashton found himself quickly absorbed into Burmah Oil's affairs; he took over from Wallace as contact man with Cohen. Despite his ability and diligence, the running of the London office remained inadequate for lack of a full-time director there.

Their opposite number in Asiatic over the agreements was Robert Waley Cohen. Although still under 39, he became a director of both Shell and Asiatic in 1906 and for more than two decades was to be literally in almost daily contact with Burmah Oil. It is difficult, from the surviving records, to decide how far he was negotiating on his own behalf, and how far he was the voice of his managing director, Henri Deterding. The impression from the Burmah Oil side was that Deterding had the entrepreneurial mind, whereas Cohen was the very gifted and astute manager. Deterding achieved his successes by simplifying complex problems and holding to his global vision of oil. Cohen, on the other hand, was able to keep all the details of complex issues in his mind and was therefore well placed to carry out policies originated by Deterding.

For Burmah Oil, contacts with Asiatic had their advantages as well as their pains. One advantage was that inefficiencies came to light, such as the wasteful overfilling of each of its 4-gallon tins by about ¾ pint. At the same time, there were many sources of friction. For instance, each company held differing views on how to use its managing agency overseas. Cohen kept his agents, Graham & Co. of Calcutta, on a tight rein and seemed little interested in any views they happened to express. The Burmah Oil people, on the other hand, expected their men on the spot to offer informed opinions and take initiatives within the overall instructions from home. Wallace reacted angrily when Cohen tried to interfere with perfectly straightforward arrangements between their respective agencies in the east. "We must leave Rangoon and Calcutta as free a hand as possible," he told Cargill. "The Asiatic Petroleum Company have got clever people here, but we have the pull over them in brains in the east."

Despite the scope for friction, however, Cohen was so anxious to create stable markets in India that he offered agreements for other refined products as well, notably petrol now that it was becoming a highly lucrative commodity. A petrol agreement was signed in July 1906. That allowed Asiatic to buy at fixed prices the whole of the company's surplus for export; from 1908 onwards, Burmah Oil was able to supply Asiatic's total petrol requirements in India, which were sold under the Shell brand name. In 1906, too, Cohen demanded a share of the jute batching oil market. Cargill had great misgivings about that, for the trade was a dependable source of profit – nearly a quarter of total earnings in 1886 – and was the more important now that kerosene production had to be held back. Cargill did have some discreet enquiries made of Standard Oil's Calcutta office, which had had jute batching oil agreements with the company for a good ten years, to see if Asiatic could be brought in. No objection was raised, and a mutually acceptable agreement was concluded in July.

However, personal relations with Cohen were not all that smooth. As Cargill put it, almost every single meeting with him ended in the "usual prolonged argument". Cargill was also wary of what he termed the "slimness" of Cohen's methods, namely his practice of shifting his ground between meetings, going back on previous decisions and then putting forward entirely new proposals. Such tactics may have been forced on him by Deterding, sending him back for one more try; sometimes he gained a concession or two, but as often as not Burmah Oil refused to budge and rode out the subsequent storm.

While on home leave in the summer of 1907, Greenway took over from Ashton. He was a tough and experienced negotiator and, perhaps in response to his initiatives in August 1907, Asiatic accepted a revised agreement which raised the limit of Burmah Oil's kerosene production by 9 per cent in return for the latter's slightly lower profit margin.

Overall, Greenway was not very impressed with Cohen. "I have been much surprised to find how weak his knowledge and information is," he told Adamson, and believed this to be at the root of the not infrequent "mistakes" made by Cohen, who never made a written note of anything but relied on a good but by no means infallible memory.[4] Ashton, on the other hand, felt that Cohen was a very able man but allowed his office to be a one-man show. Moreover, apart from a total of several weeks spent there on two separate trips, he had no real first-hand knowledge of local conditions in India. Greenway's remark over one particular manoeuvre was fairly typical: "We shall have a tough fight with him, but if we keep firm he will probably cave in."

The need for economy in every aspect of the company's operations, to husband oil supplies as well as finance, kept the technical staff in Glasgow on their toes. In 1906, for instance, they became interested in a process, patented by James Noad and W. McMullen, which involved cracking the oil and thereby changing its chemical composition so as to increase the yield of the now more valuable lighter fractions. Yet Campbell persistently opposed the whole principle of cracking Burmese oil, because its wax and other heavy products could be damaged and devalued in the process.

Even so, in 1907, the company signed an agreement to acquire the cracking process and a company, the New Oil Refining Process Ltd, was registered with a £50,000 capital. Burmah Oil held a majority of the shares and Cargill and Hamilton were two of the four directors. Campbell unenthusiastically carried out experiments in the refinery laboratories; not until mid-1911 was he convinced that the process was well enough advanced to make it a commercial possibility. With his modified approval, the experiments were transferred in 1912 to a newly opened works at Silvertown in the east of London.

The works were closed down in 1914 because the process, one of the earliest in the world involving cracking, encountered a number of technical difficulties. In particular, at the high temperatures involved, the plant was subjected to stresses and strains of which the engineers of the day had no previous experience. By 1918, rival companies had devised more sophisticated cracking systems so that the directors of Burmah Oil were not tempted to undertake any other such initiatives.

Now that kerosene prices generally were higher, the company could face the world from a much healthier financial position than previously. By 1907 earnings in Rangoon, at £726,000, were more than double those of 1905, even after generous provisions for depreciation. Profits per barrel had likewise more than doubled from 11p to 23p. At the end of 1906 it had nearly £800,000 on deposit, and £887,000 at the end of 1907, being kept against the time when much would be required once the Persian "big company" was formed.

Yet as financial problems began to recede, future supplies of crude continued to be a nagging worry. In November 1905, as soon as the fuel oil agreement was signed, the directors wrote to the India Office stressing not only the pecuniary sacrifice they were making on the Admiralty's behalf, but also the necessity of developing and proving fresh oil-bearing lands to ensure that the contract could be fully honoured. They also complained yet again about the blocks being reserved against them.

The response in India was to send the Geological Survey's specialist in Burmese oil matters, E. H. Pascoe, to work out Burmah Oil's concessions needs. The company was assured that, meanwhile, none of the blocks for which it had applied would be granted to rival applicants. Pascoe did indeed carry out a very thorough investigation; he was the first to prepare systematic geological maps for the whole of the known oil-bearing areas. His report broadly upheld the company's case, but mistrust of the company by officials was so deep-seated that the directors could not expect any immediate remedies to their grievances.[5]

Then at the end of 1906 the worry over rivals' activities, dormant since 1903, began to erupt once again. The Rangoon Oil Company had admittedly been an irritant, but at least did not refine, and it produced only 4 per cent of the province's total crude output, all at Yenangyat. However, the following year it leased wells at Yenangyaung from the Twinzas and started to drill by machinery there. In October 1907, when it nearly trebled its issued capital to £75,000, its drillers opened up a well that gave it three times as much production overall as two years before.

Some of the heaviest speculation ever seen on the Rangoon stock exchange then followed, with Rangoon Oil shares changing hands at over twenty times their nominal value. The euphoric Mower offered Burmah Oil a large block of shares at those inflated prices; the offer was politely declined although the advice from Finlay Fleming & Co. had been to buy them as a means of gaining control of a rival.

The last thing Cargill wanted at that time was to alienate officials by acquiring a monopoly through the back door. In any case, coming to the heart of the matter, he remarked to Adamson, "I can see that with a controlling interest in the company [Rangoon Oil] we could stop reckless overdrilling, but I do not quite see how the arrangement could cheapen the cost of crude oil." His cautious views prevailed. So far from buying, Burmah Oil sold out at the top of the market all the shares it had bought in 1903, thereby reaping a handsome capital profit. A few months later, to help stabilize the market, it agreed to buy much larger quantities of crude than before from Rangoon Oil.

By then a new rival was establishing himself in the fields: Abdul Kadar Jamal, head of the Rangoon mercantile firm, Jamal Bros & Co.[6] His was a genuine story of rags to riches. As a young man he had emigrated from India to Burma where he started off with a modest stall in the Rangoon bazaar. From there he had launched out into cotton and oil, the latter after he had acquired a number of Twinza well sites. Early in 1907, he began operations with the help of a head driller formerly with Burmah Oil, which could indeed claim the dubious distinction of having been the main provider of key staff for most of its competitors.

Rapidly discovering oil in 1907, Jamal at first collaborated with Mower and Clifford in a finance company to raise cash for his investments. That collaboration alarmed the Burmah Oil directors, who reckoned the latter to be little better than a pair of crooks and mistrusted their unholy alliance with so inveterate a gambler and hyper-optimist as Jamal.

Jamal, who was known to be planning his own refinery, then made an offer to the company, via Finlay Fleming & Co., either to buy his kerosene outright or alternatively to market all his refined products. The directors at home gave short shrift to that offer. Cargill, from personal recollection, condemned the refinery site Jamal had acquired at Seikkyi, 4 miles south of Syriam on the Rangoon river, as being totally unsuitable and far too exposed to the south-west monsoon. Yet what if the new refinery did make good and the consequent flow of products into Burma and India disrupted the orderly market Burmah Oil and Asiatic had been at such pains to create?

Cargill was so haunted by the prospect that he enquired of Finlay, in March 1908, "Is there any hope of buying Jamal Bros outright?" Jamal and his associates refused, and Cargill remarked rather testily, "Let them stew in their own juice," the same phrase as Kirkman Finlay had used over Minbu Oil a decade earlier. As soon as his refinery was about to come on stream, Jamal raised the stakes by threatening to throw all his refined products on to the market if he could not sell them to Burmah Oil. Yet for all his self-assurance, he was known to be desperately short of money. Finlay judged him to be "squeezable", the future being "not entirely without risk" for Burmah Oil should its directors refuse him terms.

Then Cohen heard about Jamal and his activities, and began to ask questions. Cargill virtually told him to mind his own business. So Cohen instructed the Asiatic agents in Rangoon to offer Jamal a marketing agreement for kerosene. That impulsive act was extremely unfortunate, as it gave Jamal the notion that he would have Asiatic to fall back on. Although Cohen hastily withdrew the offer and proclaimed that on this issue Burmah Oil and Asiatic were of one mind, the damage had been done. Jamal felt encouraged to look further afield, and was even reported to be flirting with Standard Oil.

For Cargill, April was a particularly cruel month, with reports arriving of the Irrawaddy Flotilla Company steamers being grounded on their way down to Rangoon – for the pipeline was not yet complete – and the Syriam refinery therefore lacking sufficient crude to meet normal demand. There was stalemate with Jamal, who declared that he would arrange his own marketing through outside merchants such as the Bombay Burmah Trading Corporation.

Later that same month Fleming was in his club when James Duncan, London head of the London and Rangoon merchants Steel Bros & Co., "rushed at" him to say that his firm was to advance Jamal the equivalent of £80,000 on the security of certain Twinza wells and some non-oil assets. Duncan was regarded by the rather staid Glasgow directors as an impossible man, whom they did not entirely trust; he now claimed that his object was merely to "shackle Jamal", who would then "take a back seat with his swelled head". When, shortly afterwards,

Steel Bros converted its projected loan into a straight purchase of half Jamal's entire business for the equivalent of £100,000 in cash, Duncan changed his tune and spread it abroad that his plan was to "wipe the eye of the Burmah Oil Company". Meanwhile, his tale to Greenway was, "We are not greedy. We only want some of the crumbs!" Cargill took that to mean that he would soon be approaching Burmah Oil, so as to unload Jamal's oil holdings at a profit.

Cargill should have known better; fortified by the extra cash, Jamal went ahead and arranged for the Irrawaddy Flotilla Company – no longer well disposed towards Burmah Oil now that its freight monopoly was about to be extinguished with the completion of the pipeline – to ship the oil to Rangoon. He also doubled the capacity of his refinery, made enquiries about possible tankers for charter, and bought sites for tank installations on the mainland of India. All this activity came to Cohen's ears and greatly unsettled him; he was reported to be in a "highly nervous" state and therefore liable to act irrationally.

Then in May, Mower and Clifford established the Rangoon Refinery Company Ltd, which acquired 125 acres of land at Thilawa, a little further south from Seikkyi. Despite the crude oil agreement with Burmah Oil being still in force, very shortly they would be refining some of the Rangoon Oil crude and thus be competing over products. This was a highly worrying prospect for the company, which would, for the first time, find itself confronting two serious rivals at once.

As if that competition was not enough, there were problems in the oilfields as well. The company's resident geologist, Basil Macrorie, wrote a strong letter to Redwood late in 1907, complaining that officials in Rangoon, who bore none of the exploration costs, knew little about the needs of the industry and had scant regard for its future. In particular, they laid down no regulations about the space between each well. Traditionally, the Twinzas had dug theirs at least 60 feet apart, a tolerable distance when depths were not great, but entirely inadequate now that machines could drill down to many times the former depths. In his view there should be at least 600 feet between the drilled wells; yet such a limit would be almost impossible to enforce because rivals had "rushed down wells" and the company had been compelled to follow suit.

On seeing Macrorie's letter, Cargill commented to Redwood, "I have no doubt this has given you a better idea than you have yet had of the extremely biassed and unfair attitude taken up by the government of Burma against us." Redwood therefore prepared a memorandum, pointing out that the arrival in the fields of new companies, with their disruptive policies for drilling, threatened Burmah Oil's ability to honour the fuel oil contract; he also mentioned the question of distances between wells.

Redwood no longer had the forum of the Oil Fuel Committee for expressing his views. It had been disbanded in 1906 after the change of government; but he sent a copy of the memorandum to Pretyman who, although now in Opposition, agreed to forward it to the Admiralty. Their Lordships noted Redwood's points and asked the Indian authorities to arrange to regulate drilling for oil on a scientific basis. That would protect the large operator, working on a "thoroughly organized basis", from being hampered by "more or less small haphazard and speculative exploiters".[7]

For once the government of India moved with commendable speed. Its Chief Inspector of Explosions carried out an enquiry, predictably concentrating on the safety aspect, and found plenty of evidence that the unregulated activities of companies in the fields were a source of "real and immediate danger". Only Burmah Oil troubled to sink wells at safe distances from one another, and the activities of others had in consequence led to quite widespread fires. Derricks were still made of wood rather than steel, and could therefore be set on fire by the nearby oil-fuelled pumping engines; and the bulk storage tanks were often too close to wells. Moreover, local district officers had no technical background and rarely stayed long enough in their posts to acquire much useful knowledge of oil affairs, thus being valueless as advisers. It was suggested that a permanent and professional regulatory body should therefore be created.[8]

The government of India at once set up a new committee with wider terms of reference. It was chaired by Sir Thomas Holland, director of the Geological Survey of India.[9] The committee questioned witnesses in depth, the Burmah Oil delegation being led by Finlay, to whom Cargill had given complete discretion on what to say. As the proceedings were *in camera*, Finlay was able to answer with absolute candour.

Holland enquired closely into the activities of the company, as the largest producer in India. To the accusation that it was overdrilling the fields in order to keep its refineries in full operation and also to deny crude to rivals, Finlay replied that the refineries had been extended long before the newcomers had appeared, and that the company was still holding part of the Twingon area in reserve instead of drilling there. As to "offsetting" wells, he admitted that he had authorized that practice, but only after ample warning. However, if the government of India would only agree to raise the tariff when necessary and also fix the distances between wells, he would be prepared to open negotiations about an exchange of sites between the firms and an agreed policy on the number of wells each party could drill.

Effectively, Holland was carrying out an inquest into Burmah Oil's performance since 1886, and he therefore asked also about the Asiatic agreement. He surmised that it was "not a bond of friendship so much as a deed of separation" (see Chapter VI above) and suggested to Finlay, "You naturally do not wish to break the agreement, as ultimate victory would be expensive."[10] Finlay did not challenge this interpretation, but put it in context by discussing the risks of a new price war arising from even relatively small newcomers like Jamal beginning to market kerosene in India.

Despite the wide-ranging scope of the enquiry, the conclusions in Holland's reports were very narrow. Strict fire regulations must be introduced: these and other measures designed to ensure good oilfield practices should be enforced by appointing a senior civil servant to the post of warden of the oilfields, assisted by an advisory board.

The government of India accepted the recommendations, but painstakingly rebutted Burmah Oil's long-standing grievances. It also dismissed out of hand Finlay's proposal for the tariff to be increased sharply enough to exclude foreign kerosene from the Indian market and thereby permit the indigenous firms in Burma to act jointly to develop oil reserves in a rational manner.[11]

The directors at home were, in turn, deeply disappointed that what had been designed as a wide-ranging investigation should end up with little more than proposals to do with fire precautions. A warden and advisory board were duly installed in the fields and did, in the event, manage to build up a creditable amount of inter-company co-operation in overcoming, for instance, the widespread flooding of wells and other malpractices.

Since the governments of India and Burma appeared to be no friends of the company, Cargill and his colleagues were all the more anxious to keep on good terms with officials in Whitehall. Regrettably, relations with the Admiralty were no longer what they had been. Gordon Miller was dead and Pretyman on the opposition benches. There was no contact with the new Liberal First Lord or his junior ministers, who were in any case committed to strict economy over defence, including oil supplies.

Moreover, according to Redwood – who passed on the news to Wallace and Fleming in the strictest confidence – the professional advisers were acting as a drag on the political heads over plans for oil, advising them above all not to commit themselves too deeply to fuel oil, as in their opinion the next few years would probably see significant developments in internal combustion engines worked by coal gas. Both Redwood and the heads of the Admiralty were sceptical about such developments taking place; but it was necessary to live with the erratic Sir John Fisher, now First Sea Lord, who was prodding the technical staff for all his worth.

Mercifully, the India Office was sympathetic towards the company. Soon after returning from the East in 1906 Cargill had interviewed (Sir) Thomas Holderness, secretary of the commerce department there and a knowledgeable ally as a former Indian Civil Servant. Cargill was seeking the explicit protection of the government of India to counter what he called "the immense resources of the Standard Oil Company of America, its unscrupulous methods of conducting business, and its willingness to spend large sums of money, if by doing so it can gain control of rival interests". Holderness, while making no definite promises, accepted the reality of the threat. Yet in common with many others in Whitehall, he was puzzled by the relative ease with which the Burmah Oil directors came to agreements with rivals who appeared to be sworn enemies.

In September 1906 the company, via Shaw Wallace & Co.,

entered into a verbal understanding with Standard Oil in which the latter agreed to limit its kerosene sales in India. Slips of paper, neither signed nor dated, were exchanged so that "the future holds no chance of misunderstanding as to what actually transpired". Nothing was specified about prices.

The historians of Standard Oil, having little on the record about such agreements, assumed them to be only temporary and subject to termination at twenty-four hours' notice. That had its uses at times: just a year later, in September 1907, its head office in New York sent out instructions to all overseas offices that every outstanding agreement must be withdrawn forthwith. A Federal anti-trust lawsuit was about to open, and it needed to be able to say that it had no restrictive agreements with other producers anywhere in the world. How soon former understandings were restored is not known.

By May 1908 Standard Oil seemed ready to revive the price war: presumably one of the "unscrupulous methods" that Cargill had mentioned to Holderness. Deterding, on a trip to New York, reported that the Americans were adopting a very firm and decidedly uncompromising attitude in striving for an increased share of the Indian market. He therefore expected Burmah Oil to reduce its shipments to accommodate Standard Oil, something which Cargill angrily refused to do. Cohen, as the man in the middle, seemed to be a good deal perturbed that Standard Oil meant mischief in India and that hostilities would soon break out in that market. Sure enough, a new price war erupted between Asiatic and Standard Oil in March 1909. Its course will be narrated in Chapter XI.

Kirkman Finlay had arranged for the pipeline route to be surveyed as early as 1902, but not until October 1906 did Cargill submit plans to the board. Not surprisingly the opposition of the Irrawaddy Flotilla Company remained unbending, Innes arguing very powerfully against the pipeline. Yet he was up against the solid opposition of Cargill, Wallace and Hamilton, and the proposal went through. Part of the cost, estimated at £700,000, was to come from retained profits and part from a rights issue of preference shares.

The laying of the pipeline began in July 1907. Most of the equipment was American, but seventeen British engineers were engaged under the direction of Charles Ritchie of the engineering firm C. & J. Weir. Its diameter was 10 inches and it was one of the earliest to use steel pipe instead of cast or wrought iron. The lengths were screwed together with couplings. Despite the risk of corrosion, most of the line was laid underground, but when it had to pass through paddy fields or waterlogged areas it rested on wooden trestles. The first section of 100 miles was completed early the following year.

As luck would have it, that year of 1908 turned out to be the worst on record for water navigation down the Irrawaddy. Shipments were so badly delayed that stocks of crude at Rangoon, normally the best part of a year's output, were down to a mere fortnight. By then Finlay Fleming & Co. were cabling home that matters had just about reached breaking point, with four benches of Syriam shut down. The company remonstrated very sharply with the Irrawaddy Flotilla Company for its lack of efficiency, and indignantly rejected counter-accusations that oil cargoes had been discharged too slowly at their destination.

Shipments picked up later in the year; then the autumn rains in the Upper Burma were the heaviest for sixty years and caused very serious flooding downstream. Thus the oil could not be pumped the full distance along the pipeline until February 1909. It travelled at something less than 1 mile an hour, taking nearly fifteen days to cover the entire 275 miles. Rather shortsightedly, only two pumping stations were originally planned; after 1918, when throughput was nearly 50 per cent greater, two extra pumping stations had to be installed. Cargill claimed that the pipeline was "one of the largest engineering works of the kind ever undertaken," which had to be carried through under climatic and other conditions without precedent in that type of work.[12] It was useful preparation for the even more arduous pipelaying operations in Persia a few years later.

In May 1909 John Innes announced his resignation from the board. Although the pipeline decision finally made up his mind, he had never fitted comfortably into the structure of personal interrelationships, which allowed the company's many problems to be dealt with fairly smoothly. To the end he remained too impetuous and talkative. In his place W. K. D'Arcy was elected a director.

D'Arcy understood very well the conventions of the business: Cargill and Wallace and, to a lesser extent, Fleming and Hamilton were the real decision-makers. The non-executive directors, including those on the London board, were from time to time consulted and were usually informed about what decisions had to be made. Sometimes they formally ratified those decisions, but otherwise were merely told of them. Yet the system seemed to be a workable one and, although individual shareholders sometimes tormented the board with endless correspondence, the shareholding body on the whole accepted the conventions and gave Cargill no further trouble for the rest of his long reign as chairman.

Notes

1 At this time Finlay Fleming & Co. were largely owned by Milne & Co. of Glasgow, which really meant John Cargill. It must be remembered that Cargill was also concerned with the agency's non-oil (and less profitable) activities, in e.g. piece goods, hardware and the selling of insurance.

2 See in particular *Glasgow Herald*, 26 and 30 April, 1 May, 3, 5 and 6 July, 31 August, 7 September and 14 November 1906.

3 Townend, *History of Shaw Wallace*, p. 43.

4 Henriques, *Sir Robert Waley Cohen*, p. 182: "Bob [Waley Cohen] never made a note of anything."

5 IOL L/E/7/545, 2903/05: Cargill to Under-Secretary, India Office, 10 November 1905 and subsequent correspondence. Pascoe's reports are dated 12 April, 2 May and 10 July 1906.

6 For Jamal and the company he subsequently launched see "The Indo–Burma Petroleum Co. Ltd" *Steel Bros. House Magazine*, June, 1955, pp. 21–8.

7 Memorandum (Confidential) by Admiralty, enclosed in India Office to Governor-General of India, 6 March 1908, IOL R&S 438/08.

8 IOL P 7967, India Geology and Minerals Proceedings, 1908, p. 569: report of Chief Inspector of Explosions, 16 May 1908.

9 IOL L/E/7/642, 1023/09 contains the confidential report of the Holland committee, 31 October 1908. Appendix II, pp. 79–129, prints the evidence of Finlay and others.

10 Ibid., Appendix II, p. 126.

11 IOL L/E/7/642, 1023/09: Government of India, Department of Commerce & Industry to India Office, 15 April 1909.

12 AGMs 1909 and 1910, *Glasgow Herald*, 8 May 1909 and 5 May 1910.

CHAPTER IX

Persia
1907–9

By the beginning of 1907, with the two wells at Shardin about 300 and 100 feet feet down respectively, the funds of the Concessions Syndicate were disturbingly low, more than £58,000 of the allotted £70,000 having been spent. What alarmed Cargill as much as the financial drain was a recurrent series of disputes with the Bakhtiari tribesmen that had been simmering ever since mid-1906. On top of that, a severe outbreak of political unrest occurred in Teheran, with the shah at odds with his Majlis, or parliament. A particular area of conflict was the exploration concession, which had been granted by the shah personally. The Majlis, angry at what they alleged to be the exceptional leniency of the terms, prepared to set up a committee to examine them, and ordered the First Exploitation Company to provide copies of the 1905 agreement and of Burmah Oil's Articles of Association.

Although the foreign secretary, Sir Edward Grey, when questioned in the House of Commons, correctly forecast that the enquiry by the Majlis would come to nothing, Cargill feared that providing extra funds for Persia was simply throwing good money after bad. In February Wallace tried to rally him, arguing with his usual diamond-hard logic that "if the whole thing were to go smash," that would be the best possible outcome for Burmah Oil. No other capitalist had the slightest hope of being able to take it over, he explained; thus Burmah Oil could sit tight and put the onus on D'Arcy to save the concession.

That was not the sort of fighting talk the despondent Cargill wanted to hear. Instead of indulging in rather conjectural games of bluff with D'Arcy, he longed for specific assurances from Whitehall that the operations in Persia would not be put at risk by further political disintegration, which was clearly encouraging the Bakhtiaris to step up their demands on the Concessions Syndicate.

Hamilton was therefore despatched from Glasgow to present an ultimatum to D'Arcy. The Concessions Syndicate would close down all operations unless D'Arcy obtained a guarantee from the Foreign Office of government protection against "trouble caused by and the demands of the tribes in Persia" or, failing that, of full compensation from Teheran. Moreover, the Concessions Syndicate would require him to bear half of of all future expenditure.

D'Arcy, already distraught by the turn of events, nearly exploded at these demands and hit out at what he saw as the Concessions Syndicate's gross miscalculation of the funds required. However, he calmed down and promised to think about it. He later unburdened himself over lunch to Wallace who was able to act as a disinterested party since he no longer involved himself in Glasgow matters. To seek a cast-iron guarantee from the Foreign Office, D'Arcy maintained, was "obviously futile to any man who understands the course of diplomacy". Wallace tactfully agreed that "no foreign office in the world would give assurances in such bold terms." When these arguments were relayed to Hamilton, he rather desperately tried to square the circle. "We of course are quite aware that we will not get a written assurance from the Foreign Office, but what we want is a definite assurance, not simply 'we will do the best for you'." D'Arcy did not waste officials' time by approaching them.

No doubt the sage Wallace had a quiet word with Cargill, for late in February the Burmah Oil board offered to put up the whole of the extra £30,000, to the limit of the Concessions Syndicate's £100,000 capital. The timing of this financial crisis was especially piquant as, in mid-December 1906 while visiting the other site of Masjid-i-Sulaiman, Reynolds had for the first time spotted that the oil shows there appeared to be far more extensive than hitherto realized. His description in his next letter home, bald and unsupported by maps as it was, much impressed Redwood, who enthusiastically hailed it as the most important find in Persia to date.

Yet in the spring of 1907 any euphoria was driven out of Reynolds's mind by the dozens of matters competing for his attention. "You will please understand," he wrote home, "that all these visits and worry with workmen and tribesmen take up time, and I find it difficult to do what I would wish and visit and examine some sections of rocks exposed in the neighbourhood." No doubt on Redwood's prompting, the people at home even commended his efforts to resist the Bakhtiaris' extortions, offering to back him to the hilt. At last Hamilton and Reynolds were speaking the same language.

For Reynolds, the overriding need now was to prove the Shardin field. As the two wells there went further and further down and the thermometer went up and up, he replied in May 1907 to Hamilton's question "when do you expect to strike oil?" He said he found the question exceedingly difficult to answer, freely admitting that his caution over Shardin contradicted the views of earlier geologists whom he regarded as having been altogether too optimistic about that field.

Sir Boverton Redwood – he had been knighted in 1905 – also wrote to ask when oil was likely to be struck, and with great tact suggested that the Glasgow directors were "fully entitled to more information than they have hitherto obtained". Reynolds took the hint well and in a full reply showed that, having failed to strike oil near the surface, he saw little prospect of locating any before penetrating right through the gypsum layer. He could not resist slipping in a final remonstrance:

> I cannot do for the company any better than I am doing now and have been doing. . . . I can well understand that Glasgow will be thinking we are doing nothing when I have it from a man recently in Bombay that their Mr Parsons who gave out that he had been sent to report on us here, reported in Bombay that we were "doing nothing and living like fighting cocks," descriptive touches which rank with the famous efforts of Ananias [the proverbial liar in the *Acts of the Apostles* who came to a sticky end].

Not until September, when the drillers were down to more than 1,500 feet in the two wells, with no traces of oil, did Reynolds find time to visit the spot where the depth of the gypsum beds could be estimated from surface indications. He calculated the thickness to be 4,500 feet; thus at the existing rates of drilling it might take many months to penetrate them, even then with no guarantee of success. Regretfully, he cabled home that it seemed quite pointless to go on drilling at Shardin, with no locality in the immediate neighbourhood at all likely to yield more favourable results.

Glasgow at once gave orders to shut down the wells for the time being and sent off to Persia one of Burmah Oil's consultant geologists, Cunningham Craig, who had been about to leave for Rangoon.[1] He spent five weeks travelling round with Reynolds; since he possessed a richly deserved reputation for being conceited, over-confident and disdainful of other people's opinions, Reynolds's forbearance in putting up with him was truly superhuman. Craig proved loftily dismissive about prospects at Masjid-i-Sulaiman, where he believed some oil might be found near the surface but little at any depth. However, he concurred that any further drilling at Shardin was out of the question. Reynolds therefore had the casings drawn from the two dry wells.

In a real sense, those tiny and remote well sites were by then as crucial a factor in world politics as D'Arcy's concessions had been in 1903–4. In those earlier years a sell-out to the Russian authorities could well have given them a foothold on the Persian Gulf. Now the set-back at Shardin could hardly have come at a worse time for Britain's foreign policy. In the previous August an Anglo–Russian convention had been signed in an attempt to clear up the areas of dispute in Asia between the two countries. Persia was to be divided into spheres of influence in which Britain and Russia, with each other's consent, would enjoy exclusive rights over all political and commercial concessions within its own sphere. This represented the third leg of what was to be known as the triple entente between Britain, Russia and France.

Whitehall was concerned because the oil workings were in the neutral zone comprising the middle of the country; it had hoped that a really thriving British oil industry in that zone would effectively block any efforts by the Russians to extend their interests towards the Gulf. A power vacuum following a failure to discover oil might precipitate the kind of dispute that could threaten the whole convention and hence Britain's defences against the emerging enemy, Germany. As a small gesture of support, in November Whitehall instructed the government of India to strengthen the consular guard at Ahwaz with an extra detachment of twenty men, who would in fact be stationed in the area being tested.

One of the officers chosen was Lieutenant A. T. (later Sir Arnold) Wilson of the Indian army, who was also appointed as assistant to the local vice-consul. This young and diligent officer, already fluent in Persian, seems to have spent most of his spare time exploring the countryside, making maps and collecting information which he later incorporated in a seven-volume official gazeteer of south-west Persia. Considering the apparent importance of the guard at a time when the Bakhtiaris were still causing bother, Reynolds surprisingly did not report its arrival to Glasgow. Perhaps he initially detested the presence of someone like Wilson as an outsider who might well be reporting back on him.

Hamilton happened to be about to leave on a trip to India and Burma, his first-ever visit to the east, and Redwood was anxious that Reynolds should meet him in Bombay so as to clear the air in a personal discussion. However Reynolds declined to see Hamilton on the grounds that he could not leave Craig, and besides had to be on hand as the Bakhtiaris were yet again threatening trouble. There was in any case, he insisted, no need for them to meet in order to patch things up; serious work was in the offing and the era of aggravation was over.

Oddly enough, no survey was carried out at Masjid-i-Sulaiman before drilling took place and Craig had to draw up his report without the aid of a topographical map. The chief assistant, Bradshaw, had been too busy making the road. On taking charge there, once a proper road was ready at the end of August 1907, Reynolds was able to choose the first well site from the many oil shows. Bradshaw and the other assistant then had to start moving plant and equipment up from the various base camps. Although the distance on the map was a few dozen miles, the contours and the twists and turns in the road made the transportation job exceptionally hard going. They had in fact just two months to complete the conveyance of all necessary equipment before the November rains began and the track that crossed the undulating Tembi river was washed away no less than fourteen times.

The convoys of waggons and mules successfully shuttled to and fro, Reynolds later paying a just tribute to the efforts of the two assistants, who took one day on their journey out, one day unloading and one day on the return journey. The climate was no help whatever; he described it as the most trying he had ever experienced, with hot winds blowing that brought no relief but merely lodged sand everywhere. That led to a general epidemic of "seediness", especially among the Europeans. Although the drilling rig for No. 1 well was assembled before the end of October, it could not be operated until they could lay on the essential water supply for the steam engine. Then the rethreading of the damaged casings was held up until a serviceable lathe could be installed. At long last, towards the end of January 1908, No. 1 well was spudded in.

Meanwhile the proposed site of No. 2 well, chosen by Craig, was being cleared, and all was ready for spudding in about six weeks later. Almost immediately both well-holes began to fill up with water, and the drillers could no longer use ropes as the moisture would have caused them to fray quickly. The alternative of drilling with rods was a much slower affair since the rods tended to fracture. The going became no easier until enough casing could be inserted to shut out the water, so the drillers went back to ropes again.

Then in April, another serious row about money erupted between the Burmah Oil directors and D'Arcy. The extra £30,000 sanctioned in the previous year was practically exhausted, and Hamilton had to tell him that this time he must bear half the future expenditure. Not a great deal of money was involved: Cargill estimated that a further £20,000 would be enough to carry the wells down to a depth that would prove conclusively whether or not oil was present "in paying quantities". Even so, D'Arcy refused and on the 15th Cargill took the matter to the Burmah Oil board. The directors gave D'Arcy until the end of the month to change his mind; if he did not, they would then consider whether to terminate the operations.

That put D'Arcy in a real quandary. To raise £10,000 was beyond him; yet, as he told his legal adviser, he could scarcely believe that 'the Burmah people would stop now." He had recently been told by James Thompson, Cargill's solicitor cousin, that the Burmah Oil Company, for its own protection, had either to prove Persia a success or show that it would be no good to anyone else. D'Arcy therefore cannily inferred that the company would never suspend drilling at a time when it seemed not too far from success, and felt he could afford to resist its demands.

Once the deadline of 30 April passed with no word from D'Arcy, Hamilton sent Reynolds instructions by letter about the future of operations, bearing in mind that his contract, already extended for six months until August, was running out. The letter was despatched on 14 May:

> We would like if possible to put the two wells at Masjid-i-Sulaiman down to 1500/1600 feet, and if no oil is found at this depth, to abandon operations, close down, and bring as much of the plant as is possible down to Mohammerah. . . . If you feel that Mr Bradshaw is capable of taking charge until the close of operations and the removal of plant, etc. to Mohammerah, kindly allow him to do so. . . .

> With regard to packing of the plant, you will arrange same to stand a voyage to Burma. Is it possible to charter a small vessel to carry it direct there?

The text was not shown to D'Arcy beforehand, but a few days later Hamilton had two difficult meetings with him and his advisers. Hamilton was convinced that the time had come to call their bluff. "If they will not listen to reason," he informed Cargill, "we will give them notice that we will stop operations." On the second occasion, "at one time it looked very like declaring war and withdrawing ambassadors." Yet Hamilton halted on the brink; he showed D'Arcy a copy of the letter to Reynolds and got him to agree that the venture should be wound up. On 19 May the Burmah Oil board consented to lend up to a further £40,000, making a total of £140,000. D'Arcy, the clear victor in this battle of wills, escaped scot-free.

Meanwhile, in Persia morale was astonishingly high despite the onset of the hot season: one that, when it was safely over, Dr Young was to describe vividly as "this last awful summer, which could have had nothing more than a paper wall between it and hell." Even the dogged Reynolds admitted, "we are all feeling it" when on 24 May the temperature shot up beyond 110°F in the shade. Yet there was almost an end-of-term atmosphere about the place, with Reynolds due to leave in three months, an assistant and two medical orderlies already gone and drillers being paid off now that Shardin was abandoned.

Then a whiff of success touched the sultry air. While those working on No. 2 well were penetrating a thick bank of salt, a "most distinct smell of gas" was reported in No. 1 well at about 1,000 feet. That spurred everyone to further efforts, even when a steel bit came unscrewed and three precious days were lost in fishing for it; Reynolds mildly pointed out in his letter home that there was a design fault which caused the bits to work loose.

As drilling was resumed, the smell became stronger and, in the powerful sunlight, gas could be clearly seen rising from the hole. As they all pressed on, the driller struck the hardest stratum of rock yet encountered, which did great damage to the bits. Once through the stratum, on 26 May success came at long last: the drillers suddenly struck oil at a depth of 1,180 feet. The gas pressure surged up and shot the oil 50 feet above the derrick, fortunately without causing any casualties. The valves were kept tightly closed, but oil forced itself through the well casings. The oil was pure and undiluted by water, and flowed at about 300 barrels a day.

Practical men, immersed in day-to-day problems, seldom have time to appreciate the onset of historic events. In fact, Reynolds's drillers had just uncovered the most extensive oilfield thus far known to man, had ensured the eventual formation of the "big company" and the recovery of the considerable sums spent in prospecting, and had taken the first step towards transforming the economies of the near and middle east. They also helped to safeguard Persia and the Gulf from further pressure by the Russians. But over and above supervising work at the well-head, Reynolds was busy sending off cables to Glasgow and London, one sent via F. C. Strick & Co. at Basra and the other through the Imperial Bank of Persia at Bushire. He also asked for storage tanks to be sent as quickly as possible, since the oil being forced out by the powerful gas pressure was rapidly filling up the hastily dug trench.

Whether his was the first telegram to report the momentous news to Britain later became the subject of controversy. Lieutenant Wilson, who for once was on the spot, also cabled the news to his superior, the consul-general at Bushire; it was relayed the next day to the legation at Teheran. As it happened, one of the legation staff passed it on to D'Arcy's agent there, a Persian national, who cabled the report that there had been "fountain of oil 25 yards high" but not specifying the depth. D'Arcy had been disappointed before, and said of this unofficial cable, "I am telling no one about it until I get the news confirmed."

On Monday, 1 June, nothing fresh arrived from Persia, and D'Arcy comforted himself with the notion that "all hands would be trying to control the fountain and measure the daily yield" before a cable was sent. That in fact reached Britain ahead of the Foreign Office cable, being decoded on the Tuesday. The precise wording, which gave the well depth as 1,190 instead of 1,180 feet, shows that it must have come via the bank at Bushire. On the morning of the 3rd, Cargill was able to read it out to the Burmah Oil board meeting just at the time when the Foreign Office was passing on to D'Arcy's agent in London the official news from Teheran: "The operators of your syndicate have struck oil at 1,200 feet which rises intermittently 75 feet above the level of the ground."

Which cable first brought the news to Britain would be of meagre interest but for the fact that the most recent histories of what is now British Petroleum quote extensively from Sir Arnold Wilson's account, published posthumously in 1941. Written as it was by an eminent public figure and based on his own contemporary diary, it ought to carry a great deal of credence.[2]

He claimed that Reynolds was annoyed because his principals in Britain had first heard the news from the Foreign Office, as a result of his own (Wilson's) despatch of a message. Since he had no code of his own, his text allegedly ran, "See Psalm 104 verse 15 third sentence and Psalm 114 verse 8 second sentence": "That he may bring forth . . . oil to make his face to shine", and "Which turned . . . the flint into a fountain of waters." In fact the cable Wilson did send to Bushire, as reported above, was in plain language, the official version arriving in England after that sent by Reynolds.

As it happens, we can check his posthumous account against a letter he sent his mother only a few days after the event, recounting how he had cabled the news to Delhi, Teheran and Bushire, but without referring to any cypher. Saving up, like all sensible young men, a tit-bit of human interest for the next letter to his grandmother, he told her how he had despatched a playful telegram to a friend about the strike, quoting the psalms. "The Persians of course did not understand, but Lorimer did." His friend Captain David Lorimer was vice-consul at Ahwaz: the joke failed as Lorimer was out of touch somewhere to the north, and Wilson had to send a special messenger to intercept him on his return journey.[3]

The discrepancy between the published account and what actually happened about the cables must inevitably colour the second allegation Wilson made in his book. He stated that Glasgow had telegraphed to Reynolds towards the end of April, ordering him finally and irrevocably to "cease work, dismiss the staff, dismantle everything worth the cost of transporting to the coast for reshipment, and come home." Reynolds, he continued, chose to ignore these instructions until such time as written confirmation arrived, on the grounds that the message contained one or two "minor errors of coding". He thereupon sent (Sir) Percy Cox, the consul-general at Bushire, a protest at the "shortsighted decision" the Concessions Syndicate was taking and the likely political and other consequences. In his book, Wilson quoted the diary entry as follows:

> I am tired of working here for these stay-at-home business men who in all the years they have had the concession have never once come near it. They have all the vices of absentee landlords.

> Cannot government be moved to prevent these faint-hearted merchants, masquerading in top hats as pioneers of empire, from losing what may be a great asset? Wrote Cox to this effect.

That letter to Cox has not been traced in the archives of either the Foreign Office or the India Office. Wilson's memory appears to have been at fault, the passage from the allegedly contemporary diary being written up years later. Certainly, his letters home give no hint of such feelings. Reynolds *did* on one occasion make the excuse for his own ends of an error in a cable, but that had been with the object of bundling the incompetent Parsons out of the country. Wilson *did* write a strong letter of criticism about the oil magnates at home, but that was in May 1909, accusing them quite unjustly of downright misrepresentation in the prospectus of the big company: charges which caused some embarrassed head-shaking among departmental officials in Whitehall.[4] What then of the telegram from Glasgow?

No doubt the Wilson version was influenced, at least subconsciously, by a controversy of 1927. The then chairman of Anglo–Persian, Sir John Cadman, while speaking in Glasgow of its early history, stated that the Concessions Syndicate had sent Reynolds a cable "to dismantle the drilling rig and move everything away", but that he had decided to await the letter of confirmation before taking action. Since Wilson was by that time in close touch as managing director of a subsidiary, the D'Arcy Exploration Company, Cadman must have relied uncritically on Wilson's account.[5]

R. I. Watson from the Burmah Oil side at once launched a full investigation to establish the truth. Apart from Wilson, the only survivors were Cargill and Dr Young. Cargill rather unusually took the opportunity in his speech at the annual general meeting of Burmah Oil in June 1927 "once and for all to 'lay' the apocryphal cable to Persia to the late Mr. Reynolds".[6] He affirmed that "no instructions to finally abandon operations were ever sent to Mr. Reynolds either by letter or by wire." Young, then chief medical officer of Anglo–Persian, was approached through a director, Duncan Garrow, who replied to Watson, "His recollection, and of course there is no more reliable authority, is that no such telegram of abandonment was ever sent or received." Dr Young had been a close friend of Reynolds, the latter being hardly likely to unburden himself to the 23-year-old Wilson whose brashness and self-confidence was inversely related to his judgment on civilian affairs.

Their testimony is the more valuable as the actual telegram books have disappeared. One recent author has even suggested that someone may have quietly removed the files to save the company from embarrassment.[7] That theory overlooks the fact that Reynolds always confirmed the wording of important cables in his next letter home, and six days after the great strike informed Glasgow, "I am not aware of your plans, and the instructions you say you are sending me may be modified by the fact that oil has been struck, so on receipt of them I can hardly act on them."

Although No. 1 well was gushing uncontrollably, Reynolds did not neglect his other work. A little later No. 2 began to spout of its own accord, but was kept more or less under control. Just before he departed on leave in August, he had No. 3 spudded in. Soon afterwards the new storage tanks arrived; yet for the moment the only practicable outlet for the oil was as fuel for the local traders, Lynch Bros. Even that involved overland transport by mule-drawn waggons. Hence the sooner the intended pipeline route to Mohammerah was surveyed, the better.

On his arrival in Britain during September, Reynolds was inclined to throw his weight around, proclaiming that he was a "free agent" now that his contract had expired, and loftily announcing that he was at leisure to see Hamilton at any time. When they did meet, he laid down some rather stringent terms for the renewal of his contract. The directors demurred, and D'Arcy expressed some private thoughts to his colleagues which they must have shared:

> I hold no brief for Reynolds, but at the same time I think he is a man who would be most useful to the new company, as he knows the country, is liked and trusted by the British authorities and gets on well with the natives, and all this is important. He is a man who will never by a stupid action imperil the concession.

Implying that Hamilton lacked the necessary skill to negotiate successfully, D'Arcy added that if Greenway or Wallace and he were to see Reynolds, new terms could be easily arranged. Reynolds finally had his contract renewed and by mid-December was back in Persia. The directors never expressed their recognition of what he had done. On the contrary, they seemed to show him even less confidence than in the past.

Reynolds returned to find that during his absence Andrew Campbell had been in Persia choosing the exact site for the refinery at Abadan, but had already left. He was especially annoyed because in his view Campbell had chosen the wrong location, and was further put out when the directors refused to give him the precise reasons for the choice. But he was soon as busy as ever with his multifarious tasks, armed with the usual set of instructions from Glasgow. The Burmah Oil geologist, Basil Macrorie, came out to conduct a thorough survey of Masjid-i-Sulaiman, clearly in order to secure the kind of information needed eventually for the big company's prospectus.

The directors had not forgotten the big company, but remained cautious for some months after the oil strike in case output fell off sharply. When in September, Bradshaw reported that No. 3 well had begun to flow even more profusely than the first two, D'Arcy at last brought himself to admit, "This news should settle all things, one would think." He therefore invoked the clause in the 1905 agreement requiring a new company to be formed once sufficient oil was discovered. As he was "being constantly worried by those interested" now that the news had become common knowledge, he asked Hamilton for a decision as soon as possible. The Burmah Oil directors took a month to think it over, and on 28 October gave the go-ahead. As Cargill was shortly to leave on a trip to the east, Wallace took over as principal negotiator. He was particularly exercised by what would clearly be the crux of the future company's operations: the provision of fuel oil for the navy.

Early in November he asked Redwood to raise with Pretyman, still an opposition spokesman on Admiralty affairs, the question of how much the new company would receive for its fuel oil. No doubt Pretyman was expected to take it up with the First Lord, Reginald McKenna, whom the directors had never met. In fact, a long wrangle ensued during which, on Wallace's prompting, Redwood reminded Pretyman of the unsatisfactory returns Burmah Oil had obtained from its fuel oil contract. Most interestingly, Wallace gave notice that if the Admiralty really wanted Persian oil before the big company were fully profitable, then it must expect to offer a price that would compensate that company for the alternative products it would have to forego. Otherwise the authorities would need to "arrive at some other means, by guaranteeing interest or something of that kind, of compensating the [prospective] Persian Oil Co. for making fuel oil".

He was recalling the fact that Burmah Oil had had to construct special fuel oil extensions to the Rangoon refinery, as well as storage tanks, for which the Admiralty had contributed nothing. This time the new company would be starting from scratch, with no other well-established products to bolster it up. Obviously, it could not expect to start earning dividends for some years to come, and would therefore have to raise money on debentures or preference shares. Thus an Admiralty guarantee of interest payments would avoid the prospective nightmare, that had haunted Cargill during the controversy of 1907 with D'Arcy, of Burmah Oil being saddled with hundreds of thousands of pounds of the new company's liabilities that the general public did not want. That idea of close links between the Admiralty and the company, although not followed up for the time being, was to leap into prominence again during the critical years 1912–14.

Meanwhile the new company had to be formed by the beginning of April 1909, since Burmah Oil would find itself in what Wallace called "budget troubles" unless it had some guarantee of the money it had sunk in Persia being repaid. Within this tight time-constraint Wallace, aided by Greenway, had to conclude all the agreements, draft the Articles of Association, draw up the skeleton prospectus for preference share and debenture issues and have it approved by the Stock Exchange council, acquire premises in London for the new company, arrange the public offer of shares, choose the solicitors, adjudicate between rival claims of bankers, stockbrokers and lawyers to be mentioned in the prospectus, and even see that the printers were paid.

They also had to choose future directors, nine being the maximum permitted. Here the selection of a really big name as chairman was crucial. The two front-runners were both former proconsuls of great distinction, Lord Cromer who had served for many years in Egypt, and Lord Milner who had been formerly in South Africa. Milner, although reported to be "fastidious" and not greatly enamoured of a new-fangled company, had a more solid reputation in investment circles than did Cromer. He was therefore approached through intermediaries, took a maddeningly long time to decide, and finally turned it down.

By then it was the end of January 1909 and little more than two months remained before the April deadline. With the indecisive Cargill on the other side of the world, Wallace was coming round to the idea of sounding out Strathcona, who had, after all, stood by the venture ever since 1905. Strathcona, having been drawn in originally by the Admiralty, insisted on talking it over with McKenna as First Lord, although Wallace less than tactfully hinted that McKenna knew nothing of the facts. Strathcona had his talk and was assured that the Liberal government fully supported its predecessor's high opinions of the venture. The First Lord undertook to provide any assistance that could properly be given and urged him to accept the chairmanship. That Strathcona did.

Of the eight other directors, D'Arcy was an immediate choice, as were the three Burmah Oil directors most closely concerned, Cargill, Wallace and Hamilton, Wallace being vice-chairman and managing director-designate, and also Greenway as his right-hand man. Pretyman was approached, in recognition of his services over the years. He declined, as a directorship would not have tallied with his front-bench duties in opposition.

Two others were chosen for their positions in public life. Sir Hugh Barnes had earned golden opinions of the Burmah Oil people as lieutenant-governor of Burmah in 1903–5 and since then as a friendly member of the Council of India in London. Oddly enough, the high-ups in Whitehall did not consider that his official post advising the Secretary of State, debarred him from the directorship of a company with Indian connections. Prince Francis of Teck was a brother of the Princess of Wales and by all accounts the ablest and most sparkling of his minor Hanoverian family. He had had some business experience in raising funds for the Middlesex Hospital, of which he became chairman before his death in 1910. While his name was being actively mooted, someone hit on the inspired *nom de guerre* of "Detective", a species then universally known as "tecs".

In the end, all matters involving the new company were arranged just in time. On 25 March, D'Arcy sold to Burmah Oil all his interests in the concession for just over £200,000 in cash and 170,000 fully paid Burmah Oil shares, then worth about £650,000. From then until mid-April, Wallace, Cargill on his return, and Greenway were completing the list of directors and renting an office of five or six rooms on the second floor of Winchester House in Old Broad Street, whither Burmah Oil was to move as well. They even had to scrounge some furniture from R. G. Shaw & Co.

A major headache was the printed prospectus, for everyone, however remotely concerned, seems to have raised objections or suggested amendments to one clause or another. As well as bankers and solicitors, there were the Admiralty, the Imperial Commissioner to the shah, and even Mrs D'Arcy who declared herself "very disappointed" that her husband's name was omitted from the new company's title; she failed to melt the promoters' hearts in what she termed a "last bid for fame". Then, on 14 April 1909, the Anglo–Persian Oil Company Ltd came a little breathlessly into existence.

Its authorized capital was £1,000,000 in ordinary shares, of which Lord Strathcona held £30,000 and Burmah Oil or the Concessions Syndicate the rest, and £1,000,000 of 6 per cent preference shares. Of the latter, £600,000 were issued to the public, together with an equal sum of debentures; in the event, neither issue had to be underwritten and both were very heavily oversubscribed, such was the eagerness of investors to participate in the new oil bonanza.

Why in the end was a separate company set up to work the oilfields in Persia? The Concessions Syndicate and the First Exploitation Company were both under the control of Burmah Oil, which could quite easily have been reshaped so that, for instance, the accounts and plant and stores departments for both Burma and Persia should remain in Glasgow and all other functions be carried out in an enlarged London office. With the considerable Persian oil deposits to supplement the failing supplies from Burma, the company could have developed into a world-wide rival of Shell, building up its tanker fleet and tank installations all over the globe for marketing oil products from Burma, Persia and any other country where it might start production. It would thus have gone a long way towards becoming the "Standard Oil Company of the east" to which Kirkman Finlay had aspired in 1901.

In fact, Deterding's inspired leadership had created for Asiatic something of that role in the eastern hemisphere, backed up as it was by the world-wide Shell organization; and Burmah Oil had realistically accepted its subordinate place. Apart from the fuel oil earmarked for the Admiralty, therefore, all products from Persia would have to be absorbed somehow in the already well-entrenched markets of that hemisphere: the alternative could only be a continuous and debilitating price war, and the painful process of reaching an accommodation was probably better arrived at by a new company.

A more fundamental reason, however, was that Article IX of D'Arcy's 1901 concession from the Persian government had stated that a new undertaking should be formed to develop any oilfields that might eventually be discovered. Hence the frequent references to the "big company". That interpretation was accepted by Sir Francis Palmer, then the foremost expert in British company law, who had earlier advised Burmah Oil.

His important opinion was that since the concession of 1901 had been granted to D'Arcy personally, with no provision for assignment except to a working company, he had no power to transfer his rights to any company, such as Burmah Oil, that had not been formed with the express purpose of working the concession. At the same time, that did not invalidate any contracts whereby another party should acquire the rights (as Burmah Oil had done) but with the intention of eventually forming a separate company.[8]

Perhaps the clause was inserted to make sure that Standard Oil or Shell could never become involved in Persia. Before the general meeting of May 1909, Burmah Oil shareholders were asking why their company should not itself have raised all the money needed and thus given them the entire benefit of the Persian oil. Cargill explained at that meeting that he had been forced by legal as well as financial considerations to float an entirely different company with its own public issue of shares.[9]

Yet tensions were bound to arise from the existence of separate organizations. Cargill was to find himself in the anomalous position of being a non-executive director of an organization largely owned by his own company, the real power resting in the hands of Wallace as managing director and of Greenway, neither of whom really went along with the cautious policies traditional to Burmah Oil. To be sure, they were bound to consult him over major issues as long as Anglo–Persian could not earn enough to finance its capital requirements and he was therefore the ultimate paymaster.

Thus the exact dividing line between the two companies was a dangerously imprecise one. Even if future developments had gone smoothly for Anglo-Persian, which they did not, the interlocking systems of directorates and financial ties would inevitably have strained their relationships. As it was, the dramatic sequence of events that occurred over the next fifteen years was to test their mutual goodwill to the limit.

Notes

1 For Craig see T. D. (Dewhurst), *Journal of Institute of Petroleum* 351–4, and Beeby-Thompson, *Oil Pioneer*, p. 116.
2 Sir A. Wilson, *S.W. Persia: A Political Officer's Diary 1907–1914* (quoted by Longhurst, *Adventure in Oil*, and by J. R. L. And *Suez: A Study of Britain's Greatest Trading Enterprise* (1969).
3 Wilson's telegram of Thursday, 28 May (two days after the strike) to the Resident at Bushire, relayed on Friday, 29 May to Teheran, is in PRO FO 371/497. The letter from Louis Mallet (FO) to Preece of Tuesday, 3 June is in *Fifty Years in Pictures: A Story in Pictures of the Development of the British Petroleum Group 1909–1959* (1959), p. 19. For contemporary evidence see Sir A. T. Wilson, "Letters, etc. 1903–21", vol. 2 in London Library (by permission of the Secretary), Wilson to his mother and to his grandmother, both 1 June 1908.
4 IOL L/E/7/642, 1037/09: Wilson to Major Cox, forwarded to Political Department, India Office, 14 May 1909. He privately believed that Reynolds was incapable of looking after the wide developments then taking place. Correspondence (in London Library), 1 March 1909.
5 *Petroleum Times*, LIII No. 1353 (Jubilee Number), 17 June 1949, pp. 420–1.
6 AGM 1927, *Glasgow Herald*, 11 June 1927.
7 L. Mosley, *Power Play: The Tumultuous World of Middle East Oil 1890–1973* (1973), p. 14n.
8 Sir F. Palmer's opinion (12 March 1909) is given in L. Lockhart, *Anglo–Persian Oil Co.*, p. 116.
9 AGM 1909, *Glasgow Herald*, 8 May 1909.

CHAPTER X

Working in the Company 1886–1924[1]

The story of a company is not just a story of events. It is also the story of people: of the top people who dominate the stage, and all the many people who have played their own parts in the evolution of an enterprise such as Burmah Oil. What, then, was it like to be working for such a company in the east, during the period until the early 1920s? Any reconstruction must start with the Rangoon office, through which the British employees, apart from the technical staff, invariably passed.

By 1924 the old merchant's residence of the 1870s, with its wooden penthouse, had been swept away. In its place was a fairly commodious office on the corner of Merchant Street and Phayre Street, a few minutes' stroll away from the wharves on the Rangoon river. There were three storeys, with the Burmah Oil staff on the top floor. Only the general manager, his assistant and the departmental heads, as well as the Burmese or Eurasian lady typists and the coding staff, had separate rooms on the side overlooking the river and Phayre Street. In the general office, the various departments had previously been divided by matting suspended from the ceiling, over which the young assistants, from their desks, had from time to time exchanged news, views, insults and jokes with one another.

The general manager not only ran the office but also coordinated the correspondence from its various departments that had to be sent each Saturday to Glasgow, with a copy to London. The departments dealt with fields, refinery, geological and shipping affairs respectively. His own policy letters he usually typed himself, for reasons of speed and security. His

146

relatively harried life can be illustrated from the tale of the general manager of the day, the over-conscientious Duncan Garrow, who in 1913 asked London for a dictaphone because he found himself constantly interrupted while trying to compose his weekly letter. The terse reply from London was, "Why not just lock your door?"

Early on, assistants were Scotsmen with a good basic but not higher education, and until 1904 general managers were in the same mould. That year C. K. Finlay took over as general manager and brought a new spirit of activity into the agency. From 1909 onwards assistants were recruited – by the company at home – through the Cambridge University Appointments Board, following the practice of Robert Waley Cohen who appointed Shell's overseas staff. As graduates in their early twenties, assistants were engaged on three-year (after 1921 four-year) contracts, renewable unless they turned out to be hopeless.

New assistants were expected to learn Hindustani and Burmese at least to colloquial standards. In 1914 their career prospects were improved by the creation of departmental managerships. The best men would in due course be made partners in Finlay Fleming & Co., and augment their incomes from the annual distribution of agency profits.

Until 1914 the company paid bonuses each year after declaration of its results, and also non-contributory pensions after twenty-five years' service. It then introduced an optional provident fund, which in 1919 was superseded by a provident and profit-sharing fund for all British employees at home and abroad. As Cargill told his shareholders later that year, the fund would give staff a permanent interest and share in the company's prosperity. Employer and employee both made 5 per cent contributions, and the company added an amount based on the latest dividend, up to an aggregate each year of 50 per cent of salary. That built up a capital sum growing at 5 per cent compound interest, which was enough to provide a generous pension on retirement. In 1924, the company contributed £145,000 to that fund while the 35 per cent dividend on ordinary shares cost £1,551,000.

British employees posted to Finlay Fleming & Co. enjoyed rates of pay that were among the highest in Rangoon houses.

W. G. Corfield, the first of the university-trained assistants to be engaged in 1909, recalled the brusque instructions he was given by C. K. Finlay on arrival in Rangoon: "Keep your bowels open, . . . have nothing to do with Eurasian women, . . . join the Gymkhana and golf clubs, and the Rangoon Mounted Rifles." Soon finding that his mess bill exceeded his income, he was told not to worry as the staff signed chits for all they bought and the firm paid their bills. Junior men were expected to run up debts to the company in the process of establishing "their proper place in the European society".

Since the directors at home saw the assistants' private balances, and therefore knew each employee's financial state, they periodically issued stern instructions that overdrafts must be reduced. In 1923, R. I. Watson as managing director strove to curb the spending habits of the assistants. Having much earlier himself been rebuked for maintaining a high debit balance, he was well suited to adopt the severe manner of the poacher turned game-keeper. He attacked as erroneous the notion of the company wishing its British employees in Burma to pursue an ostentatious way of life and participate to the full in all the province's sporting and social activities. On the contrary, he went on, its reputation could only be harmed if people imagined it was making such large profits "that its employees can afford to set a style of living to which other business men in the community cannot hope to rise."

As usual A. B. Ritchie, then general manager, vigorously defended his staff. It was right and proper, he argued, for their "style of comfort" to be at least equal to that enjoyed by assistants in comparable firms. So far from living extravagantly and beyond their means, they were "exceedingly quiet-living", with none in debt to outsiders. Besides, each had to find the money for a motor bicycle and possibly a pony, and make provision for home leaves every four years. They therefore spent less time in the clubs, apart perhaps from the golf club, than did assistants in rival houses. Only one or two of them belonged to Rangoon's social set: those were the keen dancers, who were naturally invited out a great deal. Ritchie concluded with spirit, "Certainly I can say that the average post-war assistant is much too independent to be influenced in his manner of living by any consideration of what he thinks the company might wish him to adopt." That was not an unreasonable conclusion when many had fought in the First World War and others had held responsible positions both in the agency and in the local defence forces. Watson dropped the matter.

As Table 4 shows, the non-British staff made up the overwhelming majority of employees in Burma. (The 1921 figures relate to all oil enterprises in the province, but it is unlikely that the proportions would have been radically different in Burmah Oil.) Only a minority of those in the refineries were Burmese, who were supposed to be too easy-going to respond well to strict factory discipline. More likely, the Burmese did not care much for unskilled work and found the low wages unattractive. However, between 1913 and 1921 their percentage in the refineries more than quadrupled from 6 to nearly 29 per cent of all races, and by 1921 the proportion of Burmese in managerial, clerical and skilled refining jobs was 29 compared with 22 per cent for other non-Europeans.

In the oilfields, on the other hand, Burmese greatly outnumbered Indians. It was natural for them to turn from helping with the hand-dug wells to becoming drilling assistants, rig-builders and coolies. Every American driller before 1920 worked a 12-hour shift on a group of wells, each having a Burmese crew personally recruited by himself but paid by the company. He started off each well, which the crew then took over while he moved on to the next. The artisans such as turners, fitters and carpenters were always Burmese or Indians; some had to be spared for Persia in the early days there. Indians were less plentiful then in the refineries as most of the labour could be drawn from nearby villages.

Cargill prided himself that the company was among the best employers of non-British staff in the east; its wages for both skilled and unskilled labour had for years been at least equal to the highest on offer, which enabled it to obtain all the labour it needed. As he emphasized at the annual general meeting of 1920, there had never been any labour trouble worth speaking about;[2] in addition to good pay, non-Europeans were provided with medical, hospital and dispensary services, and free housing in the fields for those who needed it. They received bonuses, and clerks of special merit were entitled to special rises. The company made *ex gratia* payments to long-serving Burmese and Indian employees who retired through ill-health or old age. Then, in 1925, pensions were given to all such staff after

TABLE 4

Burma: nationalities employed in oil industry

	Burmah Oil				All oil companies			
	All operations		Refinery staff		Refinery staff		Fields staff	
	1904	(%)	1913	(%)	1921	(%)	1921	(%)
European and Eurasians	205	(3)	233	(3)	741	(2)	24	(2)
Burmese	1,091	(14)	503	(6)	9,472	(29)	831	(73)
Others (almost wholly Indians)	6,334[a]	(83)	7,363[b]	(91)	22,867	(69)	279	(25)
	7,630	(100)	8,099	(100)	33,080	(100)	1,134	(100)

Notes: (a) of whom 87 were Chinese
(b) of whom 122 were Chinese and 1 was Philippino

Sources: 1904 and 1913 Burmah Oil Records; 1921 Burma Industrial Census

twenty-five years' service, the rate being about a quarter pay except for clerical staff, whose pensions were assessed on merit.

During the 1914–18 war, unrest over pay seems to have been avoided as the cost of living was kept down by the plentiful supply of the staple foodstuff, rice, which could not be exported because of shipping shortages. Once the war ended, the price of rice doubled. This led to non-European refinery workers being granted a 10 per cent increase in wages and a reduction in the working day from 10 to 8 hours; other employees received comparable increases.

However, discontent remained and in 1923 a strike broke out among non-European fields workers.[3] Watson not only rejected Ritchie's proposal to dismiss all the strikers and bring in the more amenable Indians, but became won over to training Burmese for more responsible jobs. As a first step he awarded large pay increases to all drilling and production teams and reduced their shifts from 12 to 8 hours. The head drilling assistant was given extra status, as well as quadrupled pay, by being designated first-grade Burmese driller. He gained full charge of his well which the American (and later British) driller merely supervised once drilling had begun. The company set up rice stores to keep down prices and prevent workers from being exploited by private traders.

It also made plans to open labour offices, first of all at Yenangyaung and afterwards at Chauk (Singu). These offices took over full responsibility for labour relations, and introduced record cards for all employees. This at first incurred some suspicion, being regarded as a device for keeping out trouble-makers. In fact, the new system facilitated both recruitment and promotion. Previously it was impossible to get a job unless one knew someone in the department concerned. Now recruitment was on an impartial basis within the overall objective of increasing as quickly as possible the number of Burmese in all grades of the labour force. Each employee was considered for higher jobs as these became vacant, being given special training as necessary. The old view of the Burmese as happy-go-lucky people of limited ability rapidly gave way to the realization that their astonishingly high literacy rate – by the 1920s up to 72 per cent of males – made technical education that much easier.

Whenever John Cargill as chairman visited Burma, he spent much of his time on welfare matters: in 1912, for instance, on having workers' accommodation built at Syriam, authorizing the payment of £500 for a Cargill ward for female patients at the Syriam hospital, and decreeing that a European nurse should replace the "tender mercies of hospital house boys" at the hospital in the fields. When Dr Bull – always known as "Papa" – first went to the fields as the company's medical officer is not known, but he looked after other companies' employees as well, and also the Twinzas whose complete confidence and trust he obtained.

His little wooden hospital had a consulting room, dispensary and operating theatre on the ground floor and about a dozen beds above. There was no sterilizer; only the powerful disinfectant Lysol. That and one or two outlying hospitals and dispensaries catered for several hundred British and Americans and several thousand Burmese and Indians. After Dr Bull retired in about 1920, Dr (Doc) Svensson and Dr Gordon Terry took over. Some of the nurses' names have survived: the frail-looking but resilient Sister Clark, later followed by Sisters Elizabeth Glencross and Helen Wilson; the first two Burmese nurses, of the Karen race, were called Dorothy and Minnie. When Watson as managing director made an exhaustive tour of Burma in 1924, among other welfare decisions he earmarked the former fields headquarters at Nyaunghla, once they were vacated by staff following a reorganization there, as a combined European and native hospital, the existing hospital premises being converted into a nurses' home.

This fragmentary account shows clearly that, in the changed conditions of greater national awareness after 1918, both Cargill and Watson were anxious to give Burmese employees greater responsibility as well as improved living conditions. Further improvements, notably in education, came in the later 1920s. The measures of personnel management, as shown by the recruitment and promotion policies and the labour office's establishment, have been claimed as representing very advanced and probably pioneering thinking for any industry in the east.[4]

There was another race prominent in the company's activities, although not as employees. The Chinese had a virtual monopoly of retailing, through the general merchandise stores they ran in every town and village. In Burma kerosene for domestic use was packed in cases containing two 4-gallon tins, which cost up to the maximum of Rs 3½ (40p). That was in marked contrast with India, where railway waggons were adapted for the bulk carriage of kerosene, eventually sold loose to consumers. Nowhere in the Indian empire could villagers or poor townsfolk afford as much as 4 gallons at a time, and they provided their own containers (usually empty beer bottles) for filling at the local shop.

The company's earliest retail agents for Burma were Chinese merchants in Rangoon, the most noteworthy being Lim Chin Tsong.[5] From an office in China Street, Rangoon, with its arresting telegraphic address of "Chippychop", he ran his network of oil distributors that extended over the whole of Burma.

Burmah Oil at first felt gratified in having such an energetic man in full control of the province's kerosene sales. Yet what he woefully lacked was financial stability. From 1908 onwards, after being awarded the sole agency for the company's product sales in the province, he became seriously behindhand in his remittances to Finlay Fleming & Co. for the purchases he made. By the beginning of 1911 he owed several hundred thousand pounds sterling: this greatly dismayed the directors and their accountants in Glasgow. They constantly prodded Rangoon by telegram and letter to reduce it to manageable proportions.

Every possible device was used. For instance, a statement of his outstanding balances had to be cabled home every week, and the clerical staff on the spot were instructed to find out the correct state of his accounts so that they could be monitored. But LCT, as he was known for short, proved too smart for the investigators. At every encounter he would by turns display charm, inscrutability, evasiveness, pathos and seeming candour; yet his interlocutors rarely discovered the crucial facts. Although perfectly fluent in spoken English, he could neither read nor write it and therefore kept all his accounts in the Chinese manner. The only person capable of interpreting them was his own head clerk. However, the transactions arising from all his different activities were jumbled together; since at various times these ranged from steam-ships and a match factory to rubber plantations and tin mining, it was impossible to distinguish clearly what related to what.

C. K. Finlay, as general manager until 1912, did his best to grapple with this thorny problem. His frequent letters to LCT were often strongly worded, but usually ended, "with plenty salaams" and the two men seem to have remained personal friends. The two succeeding general managers strove hard, but unavailingly, to get LCT to pay up. Only the outbreak of war saved him from a complete débâcle and Finlay Fleming & Co. from massive bad debts. Many vessels he owned were then chartered by the government and thus provided him with a risk-free income to set off against the money he was squandering elsewhere.

LCT became a member of the legislative council and donated large sums to local war charities, later being given the OBE. He also built a palatial residence in one of the smartest quarters of Rangoon, shaped like a star and furnished with mural paintings by the later Royal Academician, Dod Procter. Yet he was still up to his old tricks: in 1916, when the government was having great difficulty in enforcing maximum prices, he was the first retail agent in the Indian empire to be caught selling kerosene at above the controlled price. By 1918, Watson had formed an inflexible resolve – and, more important, had persuaded Cargill – that LCT must be fired, and replaced by a new and more trustworthy system.

As from the beginning of 1920 Finlay Fleming & Co. took control of all marketing in Burma. The province was divided into four marketing districts, each run by an assistant on the spot who was backed by a specially augmented staff in Rangoon. As all transactions were for cash, the organization worked efficiently, and Cargill's fears that he would lose business proved to be unfounded.

On the change-over date LCT, having resigned before he could be given notice, was well outside Burma. He had taken over one of his own liners, the *Seang Bee*, to fulfil a lifetime's ambition of visiting London together with his family, friends and business associates: a trip that seemed crazy enough to inspire an ironic comment in *Punch*, although in fact he handsomely recouped his expenses from the cargo the liner carried. In London he waited on Cargill and other directors, who entertained him and his party to luncheon at the Hyde Park Hotel. All business talk was carefully avoided until a formal meeting was later held in the London office.

The company sought to be generous in its final settlement with him, and they parted in complete amity. He died three years later, aged 53, of a heart attack brought on by a threat from the telephone company to disconnect him for non-payment, together with a clearly unwelcome "communication" from his bank. Watson personally mourned the death of a man who, in carefully chosen words transmitted to Rangoon, "after all was a long [time] servant of the company and at any rate, up to a period, served it faithfully and well".

Of all the company's workers, the most colourful were undoubtedly the drillers who until the 1920s were nearly all American. The nature of their difficult and demanding work made them a law unto themselves. Daniel Dull of New York, with responsibility for recruitment, had been instructed to select men who were both efficient and amenable, but later began to send out drillers who proved to be neither, so that matters got out of hand.

C. B. Jacobs, as fields manager from 1902 onwards, built himself a very lavish bungalow at the company's expense. That was an example of what C. K. Finlay later called the "damn-the-cost-system", which allegedly dissipated the company's resources and encouraged many other Americans to feather their own nests. The appointment in 1905 of a fields agent on the spot, referred to in Chapter VI, was clearly intended as a first step in securing a grip on what Finlay described as "semi-controlled Yankee management".

Drillers predictably resented the presence of the agent in their midst and when G. C. Whigham, who had a reputation of being a "hustler", took over early in 1910, his precipitancy brought them out on strike. The company could only get them back to work by replacing Whigham with the more easy-going Hubert Heath Eves: a deplorable climb down, according to one observer. All this led the directors to pin their hopes later that year in Thomas Hayes, hoping he would prove a new broom who could impose greater efficiency and economy in the fields. As will be seen in Chapter XI, these hopes were to be cruelly dashed.

To be sure, the drillers did have a difficult time, being badgered by the office staff about such peccadilloes as leaving pieces of equipment lying around the fields, when they had more important things on their minds. It was no fun hurrying round from one newly drilled well to another, especially in the hot weather, when shifts were 12 hours long. Nor, in the early days, were there any particularly congenial leisure conditions in which to unwind. The company forbade them to bring their families, although they were paid highly enough to be able to support them in the province. An equally rigid rule, dropped some time in the early twentieth century, forbade them to drink liquor.

The unruliness of the drillers became common knowledge outside Burma. When drilling staff was being recuited for the Persian oilfields in 1908, a consular official there wrote, "I have heard from various sources that the state of indiscipline [among American drillers in Burma] is disgraceful. The men drink and fight among themselves, shooting affrays being not uncommon, and they prohibit the company from employing any but American-made plant." To be sure, much of the equipment was available only in the United States, but various stratagems had at times to be used to overcome the drillers' prejudices. When they objected to British wire rope and insisted on having it ordered from America, the stores official simply wound the British rope on to an American reel, whereupon it was readily accepted.

The drillers themselves had to overcome a major lack of welfare facilities. Not being eligible for membership of the European – and predominantly British – club at Yenangyaung, they set up, independently of the company, an American club in 1909. It moved to more permanent quarters in 1914, situated in its own grounds, half a mile from the main oilfield. The single-storey building had a timber frame and, like all buildings in the area, was filled in with matting. A verandah ran all round.

Members were not allowed to enter until they had "parked" their guns at the club entrance, for fights broke out not only as a result of personal quarrels but also through territorial rivalries, Texans and Californians being particularly incompatible. Once the ban on liquor had been lifted, there was a large bar, but the most popular part of the club was the card room. At the week-end, a poker game would begin as soon as normal work ended at noon on Saturday and continue for 36 hours on end until the bar closed at midnight on Sunday. As many as ten members could sit round the table, never becoming noisy or over-drinking. A shaded kerosene lamp hung overhead, and the stake money was in a pot, in five-rupee (33p) notes and above; coins were never used.

As there was virtually nothing in the area for members to spend their money on, stakes were often extremely high. When players ran out of money they would sometimes wager the ponies they used for travelling round the fields, and later on even their motor cars. The money was seldom counted and never checked, but just added to or taken from the pot as the interminable rounds progressed. Players came and went, leaving their money knowing it would not be touched. Drinks could be ordered for as long as the game lasted: a notice above the bar read, "This bar is closed from 2 to 3 a.m. on Sundays: the bar boy has to have his breakfast some time." Monday mornings were especially uncertain times for the Burmese drilling crews. If their driller had won, he would hand out tips; but if not, they could expect blows and curses if anything went wrong.

The no-wives rule was relaxed at about the time of the First World War. This dramatically changed the club's character: it no longer resembled a set for a rather bad Hollywood western but, as always when a feminine touch is introduced, rapidly became more civilized. A new focus of interest emerged: the large sprung dance-floor, made of the finest teak. The band, comprising a pianist and a drummer, was formed from among the members themselves, while the wives organized the refreshments and meals. To make sure that the ladies were well circulated, the men always demanded plenty of Paul Joneses and "excuse-me" dances.

A wilder dare-devil spirit had not, however, been wholly extinguished. There was an occasion when a number of exuberant drillers – presumably bachelors – caused a stir by riding their ponies on to the ballroom floor. Then in the 1920s the number of Americans began to diminish as British and Burmese nationals became involved in the drilling work. Not surprisingly, there was little contact between the club, with its free and rather outlandish ways, and the far more sedate Yenangyaung club for British members, the tennis club at Khodaung, or the BOC club at Nyaunghla for the office staff.

The Yenangyaung Club became the social hub of the oilfields for British staff.[6] Subsidized by the oil companies, club membership was by election, the club's committee being itself elected by the general membership. Little remains on record about its early days but by the 1920s the club premises consisted of a bar, card-room and dance-hall, with outdoor facilities for eating and entertaining. There were hard tennis courts and a nine-hole golf course. The fairway was bare gravel and the greens, known as "browns" were constructed of oiled patches of sand, brushed back into immaculate condition immediately after use.

Informal Saturday evening dances were held most weeks, punctuated by a formal monthly dance. Sunday breakfasts were a major social event, beginning at about 11 am with gin slings or King's pegs (champagne cocktails) followed by a satisfying meal normally structured around chicken mulligatawny.

The arrival of university-trained assistants after 1909 gave a new impetus to sporting activities. Cargill's correspondence with the east not infrequently mentioned cups and other trophies which the company or he personally presented for a wide variety of sports. Football came into its own, particularly in the monsoon season. Teams were of mixed British, Burmese and Indian personnel. Each of the smaller oil companies usually fielded one team. Burmah Oil provided three, representing its head office, engineering and transport divisions. The matches were played on a league system and evoked great sporting rivalry.

For a Finlay Fleming & Co. assistant, it was a strict rule in those days that he could not marry for at least seven years after recruitment. Even then he had to obtain permission from home and prove that, from his salary and any private means, he had at least Rs1,000 a month (£67). The reason was purely financial. Many organizations at home warned their overseas employees against over-hasty marriages when too young; companies such as Burmah Oil could have found themselves heavily committed to bailing out couples overseas who found it difficult to make ends meet.

Cargill conscientiously vetted all cases and, where possible, discreetly interviewed the fiancées. He declined to stretch the

rules even if applicants' circumstances were better than average. Once, an assistant stationed at Chittagong wrote to say
that he had become engaged to a tea-planter's daughter and
would therefore be free of financial insecurity. That argument
cut no ice; when permission was refused he replied that he was
very sorry but the banns had already been called. His contract
was at once terminated, Cargill observing rather late in the day,
"He was not at all the class of man we want."

For a young man in his twenties to have to spend seven to ten
years in the tropics with relatively little chance of normal
female companionship was a deprivation indeed. Some took to
drink; by the end of the nineteenth century at least two senior
members of the Rangoon agency had become utterly dependent
on alcohol. Others cultivated various forms of eccentricity. One
accountant as late as the 1920s was nicknamed "lightning
Jack" because of the extreme deliberation with which he
performed every task. As a colleague later recalled, "his bath
after office and preparation to go to the Gym [Gymkhana Club]
lasted about 1½ hours, during which his 'boy' fanned him with
a towel like a second in a boxing ring."

The British employees in the refineries, mainly Scots and
therefore accustomed to making the best of conditions overseas,
enjoyed a more cheerful and at times rumbustious existence. A
single issue, the second and almost certainly the last, of the
Syriam Trumpet, dated 1907, has survived and gives a facetious
picture of life there. The twenty "Don'ts for Probationers"
included the following:

> Don't imagine that you will revolutionize the colony. It takes a lot
> to impress the pensioners.

> Don't advocate weekly concerts. Your repertoire is large, but it
> shall diminish in five years.

> Don't accuse your boy of laziness. He rises earlier than you, and
> retires later.

> Don't ask the boy to dress and undress you. You are not here for
> good.

> Don't attempt to impress the natives. They are only impressed by
> your obvious attempt.

> Don't acquiesce in any new theology, however suited to your
> temperament. East of Suez there *are* ten commandments.

In Coke Row, where the unmarried employees lived, "a jovial lot of chaps" would from time to time treat themselves to a late night celebration, dancing the lancers and quadrilles with one thumping the piano and another blowing some wind instrument. After those exertions they restored their energies with a hearty menu of poached eggs and cake.

Most of the assistants were from independent schools, and here was the justificaton of the public-school system at that time: the segregation and emphasis on self-reliance and on character-building as much as (and often more than) academic prowess, the running of everything that mattered by those only marginally older than the youngest, the heavy reliance on structured sport and a tightly organized day, and an almost imperceptible gradation from cowed insignificance to god-like authority. In its heyday, the empire was largely administered on that kind of training, and it proved its worth for the company when young men still in their twenties were given much responsibility over large sums of money or groups of people, often in outlying districts away from close supervision.

However, given the strict no-marriage rule, it is not surprising that employees from time to time found themselves in trouble. One was given permission to return home at short notice because of what the high-ups coyly termed his "bump of philoprogenitiveness"; it was other men's wives he loved, not discreetly but too well. The people at home thought of sending him on to the Persian refinery at Abadan, but finally decided that his "peculiar tendencies" might wreak too much havoc in the then tiny mixed community there. In 1917 a geologist had to resign in a hurry to join the Indian army for much the same reason. Yet another, after the war, was cited as a co-respondent in a Rangoon divorce suit. But London and Glasgow found it difficult to decide where he could best be placed; to complicate matters, the lady in the case was Eurasian and thus, were he to be sent to a small station such as Digboi in Assam, the couple's social position could be doubly delicate.

Predictably in the circumstances, employees sometimes struck up friendships with Burmese women. In the province, in contrast with India, there was no caste system and no purdah for women, who moved freely in every part of society. Such relationships were often based on real affection, giving a lonely man an amiable companion who could discuss the day's happenings with him and laugh him out of his worries. As a very objective author has written, "surely a man who had loved a Burmese girl must at least think of the Burmans as human beings, not as columns of figures in fortnightly returns."[7]

Occasionally, a man preferred to remain with his Burmese companion even after he was free to marry. One senior partner of Finlay Fleming & Co. who retired to Scotland as a bachelor and died there in the early 1900s, is known to have had a Burmese "wife". When an employee died in 1913, C. K. Finlay in London advised Rangoon about compensation for the widow. "If she be Burmese or Eurasian and 'pucca' [legally married]," he suggested, "I should give her say Rs3000 [£200]"; if British she was to receive the equivalent of six months' pay and a homeward passage on a Paddy Henderson steamer. The deceased was none other than the unfortunate Mr Parsons who had incurred G. B. Reynolds's wrath in Persia; mercifully he had been lately performing "good solid work for the company".

If the British were discreet in arranging any affairs with Burmese women, the American drillers were far more open. The company's earlier no-women rule at the fields had simply left a vacuum in their private lives which they filled to their own – and presumably their partners' – satisfaction. Liaisons were so prevalent that the government of Burma eventually had to pass legislation to make sure that they left the women and any children well provided for when they departed. Normally, however, they could be trusted to be generous enough, some dependants even having their own banking accounts. Even so, Burmese women were never allowed to visit the club.[8]

Only one attempt was ever made to clean up such irregularities. The wife of a very senior director of the company some time before 1914, is reported to have made a pioneering trip to the oilfields, "hitherto untroubled by memsahibs". There she was horrified to learn about the goings-on, and on returning home made such a fuss in the Glasgow office that the decree went forth to Finlay Fleming & Co. that the girls were to be sent packing. Rangoon passed on the order to the fields agent who passed it to the general fields manager, presumably C. B. Jacobs. He called together a meeting of drillers at the American club and at once came to the point. A stunned silence fell; then "the meeting exploded with wrath. Everyone spoke at once,

and very harsh sentiments were expressed." Only minutes later
the fields manager cabled Rangoon. His message was earthy,
and may be politely paraphrased as: "No girls, no oil." It was
pointless to argue any further: the girls stayed and the oil
continued to flow.

To most British employees, however, local liaisons of a serious
nature would have been unthinkable. They had engagements
or understandings with young women at home, and many
couples remained touchingly loyal to each other during the
interminable waiting period. One, W. G. Corfield, already
engaged when recruited in 1909, was unable to marry until the
end of 1915 even though he had in the meantime become
assistant general manager in Calcutta. Some fragments of a
diary for 1910, clearly kept for his fiancée, provide one or two
unforgettable vignettes of life in Burma. That year, he travelled
up-river at night to the oilfields in an Irrawaddy Flotilla
Company steamer, the powerful searchlight on at full beam to
pick out the river's treacherous bends. "When the light was
turned on to the bank," the young man wrote, "all the kiddies
came dashing down and performed the most grotesque dances
on the shore. It reminded me somewhat of the moonshine scene
in *Our Miss Gibbs*" – or no doubt quite a few other musicals
before and since.

When marriage had finally been sanctioned, a sensible wife
became totally committed to her husband's career: in a real
sense she had married not simply a man but also the company.
Conditions varied greatly according to where they were stationed; the comparative hardships of life in the fields, for
example, had to be shared together but there was usually plenty
of work to fill the time. In Rangoon, people were more grega-
rious, enjoying a wider social life among their fellow-Britons
and far more opportunities for shopping and paying visits. Not
that there was much genuine contact with the indigenous
people they lived among. The wife of a prominent character in
this book, a government official's daughter who was born and
married in Rangoon and lived on there for the best part of a
decade, was asked many years later what she had thought of the
people. "The Burmese?" she replied, "Oh, I don't think I ever
knew any."

Domestic servants were almost always Indian, and sought to

cocoon the memsahib from life's realities. She might order the meals, or could just as well leave it to the cook. To cook or do housework or "do her own bazaar" (shopping in the market) would have meant losing face, and that made bridge or golf into immensely popular pastimes. Only gardening was permitted by custom, so long as the manual work was performed by the mali or gardener. It is a tribute to the initiative and inventiveness of so many wives that they put their creative talents to devising elaborate floral displays in the gardens of their official or rented bungalows and floating candles during Thadingyut, the Burmese festival of lights.

Life became more complicated when children began to arrive, although the ayah or nurse was essential until the children went home to boarding school. It was only in the early 1920s that one or two of the unmarried assistants began to employ Burmese instead of Indian house-boys; until then everyone had assumed that the Burmese did not have the right touch for performing such personal services. In fact, they seem subsequently to have become highly efficient and devoted servants.

By the mid-1920s, the company estimated that in the Indian empire it was giving direct employment to over 40,000 and thereby supporting a total population of some 200,000. It also provided indirect employment, wholly or partially, for about half that number again.[9] It would be impossible to give a balanced account of their lives, not least because few, even among the British, seem to have troubled to write down what company life was really like in those days. The present chapter has therefore tended to supplement the bare surviving facts with some of the unusual and outlandish occurrences that tend to be remembered after the routine ones have been forgotten.

It is with the senses as much as with the mind that most expatriates will have remembered Burma. The rich, often gaudy, colours of Burmese dress; the high-pitched long wailing sounds of Burmese music which Kirkman Finlay could not have been alone in finding reminiscent of his native Scotland; the pagodas outlined against the dawn; the bitter-sweet smell of the ground cover during snipe-shooting; the prickly thorns that one's mount refused to jump while out for a gallop; and, not least, the sad awareness that all this was long ago.

Notes

1 For this chapter (and others) see R. I. Watson's "The Burmah Oil Co. Ltd.", reprinted from *Daily Telegraph*, February 1927 and issued to shareholders, 1927.
2 AGMs 1920 and 1921, *Glasgow Herald*, 25 June 1920, and 8 July 1921.
3 AGM 1923, *Glasgow Herald*, 13 June 1923.
4 P. P. Higginbotham (formerly Labour Officer Burmah Oil), "BOC's program for training Burmah technicians and administrators now is 30 years old", *The Oil Forum*, January 1954, pp. 21–5.
5 An account of Lim Chin Tsong's bizarre career (by H. H. Twist) is in *Burmah Group Magazine*, No. 6, Summer 1966, pp. 18–21.
6 Dr C. T. Barber, "Social life in the Oilfields", *Petroleum Review*, October 1974, pp. 663–6.
7 P. Woodruff, *The Man Who Ruled India: II The Guardians* (1954), pp. 129–30.
8 Beeby Thompson, *Oil Pioneer*, pp. 500–1.
9 Watson, "The Burmah Oil Co. Ltd.", p. 14.

CHAPTER XI

A Period of Transition (2)
1909–11

In the history of Burmah Oil, although not much happened internally in 1909–11, the events of those years exemplify the volatile nature of the company's external relations. Oil companies of that era had to be prepared to move with great rapidity from outright competition to agreed collaboration with rivals, not necessarily to reap abnormal profits – and certainly not in Burmah Oil's case – but because of the very nature of oil.

Few products can be both as flexible and as inflexible as oil.[1] There is nothing flexible about its geographic distribution. The principal areas of discovery have been in some of the world's most desolate deserts or, more recently, under its most stormy seas. For every producer-well, a vast number of dry holes are expensively drilled. The enormous costs of prospecting, involving long time-scales before oil is discovered and produced, are inescapable factors.

Once oil is found in commercial quantities, it is relatively easy to transport and yields a whole range of petroleum products in the refining processes. Yet this important flexibility has its built-in inflexibilities. The product range cannot be all that widely varied, even today. Inability to sell the whole barrel, which is to say all the products capable of being refined from a barrel of crude, could bring the refining process to a halt unless there were sufficient storage to accommodate unsold products.

All these factors were at least as relevant at the turn of the century as they are today. They limited the degree of competition an oil company could indulge in, and the most successful oil magnates were those who knew to a whisker when the rivalry had to stop and an accommodation sought, even with erstwhile deadly enemies. As Calouste Gulbenkian, the oil operator who made a fortune from his 5 per cent shareholdings has put it, "Oil men are like cats; you can never tell from the sound of them whether they are fighting or making love."[2] Their rapid transition from competition to collaboration and *vice versa* baffled politicians and officials who thought that oil was flexible, and who failed to understand the problems associated with its inflexibility.

As early as January 1909 the established oil relationships in India were shaken when Jamal shipped the first consignment of his own kerosene to Calcutta. The Asiatic directors, represented by A. S. Debenham, deputizing for Cohen who was laid low with appendicitis, at once called on Burmah Oil to reduce sales by the same proportions as themselves and Standard Oil in order to offset the new imports. For the company, Ashton protested that overall kerosene sales had increased by nearly 50 per cent since 1905 with very little coming its way; now that there had to be a cut-back, it was only fair for Asiatic to bear that burden.

In the absence of the nervous Cargill, out of immediate touch in Rangoon, Wallace fully backed Ashton in his firm stand. Fleming, in a rare interjection, added the pithy comment, "There will be no end to it if we are always to give way to the Asiatic Petroleum Company whenever they come whining round." Debenham therefore summoned Deterding, whereupon Wallace unceremoniously took over negotiations from Ashton.

As the brains of their respective organizations, Wallace and Deterding hammered out a workable policy towards Jamal. Logically speaking, they agreed that there were no grounds for giving him a market share, for a newcomer should not expect to dictate unreasonable terms to companies that had already spent large sums on developing the markets. Yet to be realistic, the new competition had to be faced and Wallace readily accepted Deterding's suggestion to grant Jamal a market share without curtailing Burmah Oil's fixed quota of direct sales in India.

In return, Burmah Oil allowed Jamal a sixth of the kerosene

market in Burma, equivalent to his share of production, and also agreed to buy the residue after the kerosene had been extracted, as well as the refinery plant no longer needed to turn out heavy products. Asiatic conceded that Jamal could sell surplus kerosene in both India and the Straits. Despite having made the most of a rather poor hand, Wallace felt highly dissatisfied with the outcome. As he reported pungently to Glasgow, "The Burmah Oil Company's buying a lot of machinery that it does not want, and making room for a share of Jamal's kerosene in Burma – our very best market – is more than a *quid pro quo*" for the fairly minor concessions made by Asiatic.

Oblivious to the settlement so carefully made over his head, Jamal that February established the Indo–Burma Petroleum Company Ltd, with a rupee capital equivalent to £667,000. Steel Bros, which held a 50 per cent interest, were the marketing agents in Burma. It was the London manager of Steel Bros, the bouncy Duncan, who strangled the accord at birth: when Deterding put the terms to him, he retorted that the offer of a sixth share in the Burmese market was a "joke" and that both he and Jamal "felt very much insulted by it".

Deterding with unusual patience tried to educate him in the facts of industrial life. Oil was quite different from, say, rice, which could be stored and mills closed with no great harm. Refineries had to be worked continuously if marketing networks were to be kept regularly supplied with refined products. The lesson failed to sink in, and Deterding reluctantly decided to let Jamal spin enough rope to hang himself. Meanwhile, Burmah Oil and Asiatic would try to speed the process by sharply reducing kerosene prices throughout India.

However, such shock tactics required the co-operation of the Standard Oil people. Embarrassingly, the Americans chose that precise moment, in March 1909, to terminate the current "understanding", and plunged into an all-out price war with Asiatic. There were long-standing grievances on both sides, such as increased American penetration of the market in India: yet the formation of Indo–Burma was primarily responsible for the collapse of the undoubtedly too cosy market-sharing arrangement that had existed in India since 1905. As Deterding took steps to match the very low American kerosene prices, he was known to be despatching very lengthy telegrams to Standard Oil's head office in New York, presumably about the price war, although the contents were not known.

Standard Oil's response was to send over its top business diplomat: no longer the veteran Libby but a 30-year-old rapidly rising star, Walter C. Teagle, already being groomed for the main Standard Oil board. Before Teagle could arrive in Europe, Cohen was back in harness; his recovery had been held back by physical debility arising from years of overwork, but his energies now seemed to be fully restored. He confidently forecast that the Asiatic directors could easily settle the entire question with Teagle and suggested that Burmah Oil might care to reimburse half the profit Asiatic had lost on the market share it had surrendered. Commenting sourly that Cohen seemed "unreasonably ready to meet the Standard Oil Company", Wallace was able to stall on that suggestion because of Cargill's absence in the east.

Cargill was due back home later that month, and his colleagues made concerted efforts to prevent him from being waylaid by Duncan at the Oriental Club in London where he was to stay overnight. They feared he might be misled into making some incautious remark through ignorance of the latest developments over Indo–Burma. Wallace and Hamilton managed to brief him first, and he readily accepted the justice of conceding Indo–Burma a fair market share within an overall agreement. After his arrival and that of Teagle, the stage appeared a little overcrowded as everyone of any note, Fleming and Hamilton included, took some part in the subsequent proceedings. Even so, an accord with Asiatic and Standard Oil was soon concluded.

In this general atmosphere of *détente*, Burmah Oil and Asiatic were able to settle their differences over another product, paraffin wax. An agreement had become essential to prevent a collapse of prices now that the Americans were producing a low-grade but cheap variety and Asiatic's refineries in the Dutch East Indies were stepping up both wax and candle output. Since 1908 Burmah Oil's total receipts from wax had declined by nearly a third despite having more to sell as a by-product of its fuel oil commitments.

When the question came up in mid-1909, Cargill proposed a joint marketing agency for wax. Cohen's terms for this were stringent and Cargill had to work hard to win some concessions along the way. Their agreement came into force from the beginning of 1910. It created a wax pool run by Cohen and his Burmah Oil opposite number in London, with co-ordinated prices and sales policies and the proceeds of all sales being divided in agreed proportions. Each company maintained its own distribution network, and was given incentives to keep costs of candle production as low as possible. For Burmah Oil the agreement was restrictive but, even so, its wax production rose from 2.3 per cent of throughput in 1910 to 4.2 per cent in 1921. It could also look forward to better returns from wax and candle sales.

Although the Americans had agreed to a truce in India, it could not possibly be the hoped-for lasting settlement. The instigator of the latest batch of trouble, namely Indo–Burma, steadfastly refused to be corralled into the fold. Ever since its Seikkyi refinery had come on stream in 1909, the employees there had committed all kinds of technical blunders that had sometimes caused serious accidents. Yet they had won through and were making serious inroads into kerosene sales in the mainland, so that their rivals could no longer brush them off.

Those inroads caused Cohen to be so "down in the dumps" that he did what Deterding had carefully refrained from doing: he threatened to give six months' notice to terminate the kerosene agreement unless Burmah Oil compensated him for his trade lost to Indo–Burma. If he had really made up his mind on that, Ashton rejoined, then he had better give notice at once because no one in Burmah Oil would for a moment consider any whittling down of its terms.

Cohen was further discomposed by the real or imagined "uneasiness" of Standard Oil over the allegedly aggressive marketing tactics of Burmah Oil retail agents in various parts of India. Directors who had spent years in the east were not gratified when Cohen, with only a few months' experience of India, constantly leapt to conclusions about the behaviour of native traders and was totally deaf to counter-arguments. His newly restored vigour had begun to drain away under an intolerable load of work. On top of everything else, he had been left by Deterding, about to depart on an overseas tour, with various major issues to settle. Cohen even arranged to go up to

Glasgow and argue out his grievances with Cargill; Ashton's comment was, "Unless I am very much mistaken, you will have to listen to some pretty tall talk from him."

Although once again at the end of his tether, Cohen proved far more amenable than expected when he arrived in Glasgow. As he barely mentioned kerosene, Ashton surmised that he must earlier have been simulating anger in order to see whether any improved terms could be wrung from Burmah Oil. "It is not the first time," he remarked of Cohen, "that I have had reason to think he was trying to pump me." Indeed, the only real business done at the Glasgow meeting was for Cargill to extract some small but welcome amendments to the proposed wax agreement.

During that autumn Burmah Oil and Asiatic jointly kept kerosene prices very low as they strove to inflict a mortal blow on Indo–Burma. That company was, according to the estimates of Finlay Fleming & Co., losing nearly 1½d (0.6p) on every gallon it sold, but seemed blithely unaware of the fact because it did not write off any of its drilling and fields costs to revenue, as Burmah Oil did. In the coming year it expected to start shipping some good quality jute batching oil to India, which could only bring further gloom and despondency to its rivals. So when in November, Jamal made some peace overtures to Burmah Oil via Finlay, no one knew whether he was prompted by financial embarrassment or by the hope of gaining a tactical advantage from his triumphs to date. In the event, Duncan of Steel Bros denied the company the satisfaction of turning Jamal down by making it clear that the latter was negotiating "without authority".

The turn of the year brought sombre news from Rangoon about profits earned there in 1909, the lowest for several years. Poor returns from kerosene and wax were mainly responsible and, as the rate war against Indo–Burma was making no headway whatsoever, Burmah Oil and Asiatic decided to call it off. Then in February 1910 Jamal made a startling offer: to dispose of his own half-share in Indo–Burma, valued at the equivalent of £250,000, in return for 250,000 Burmah Oil ordinary shares at par; each £1 share then had a market value of about £5. He also demanded a seat on the Glasgow board and gave the company four days in which to reach a decision.

Finlay, when cabling the terms under the strictest secrecy, dismissed them as ridiculous, an epithet heartily echoed by the directors who categorically rejected them. Jamal was content to sit back for the moment, for one of his newly drilled wells was giving such good yields as to double his total output of oil. Meanwhile, rumours to which Finlay gave no credence indicated that other parties, including Steel Bros with their existing 50 per cent stake, were after his oil interest.

Nothing if not persistent, Jamal in June renewed his offer to Burmah Oil, this time demanding £1,500,000 in cash for his half-share: Greenway calculated that even allowing for the higher output of crude from the new producing well, he had overvalued his oil properties by about 50 per cent. Yet, according to reports from Rangoon, Jamal's drilling, refining and installation costs so outran his total resources that Steel Bros would eventually have to bail him out. Whatever the truth, Jamal was regularly in and out of the offices of Finlay Fleming & Co. to see if a cable had yet arrived accepting his terms, and to pledge full co-operation – unofficially, of course – once the company declared a truce with him.

Jamal might possibly have believed that Burmah Oil would come to terms because of some ominous moves on Mower's part. Mower merged Rangoon Oil with the Rangoon Refinery Company, whose works came on stream in August 1910, into a new combine registered that month in London as the British Burmah Petroleum Company Ltd. Its issued shares – £1,900,000 – were identical in value with those of Burmah Oil, but they were not for public sale and were to compensate the vendors as part of the purchase price.

Familiar as they were with Mower and his shady business methods, the Burmah Oil directors had a lively distrust of both the new company and its top management. The prospectus evoked loud cluckings of disapproval; it was "grossly misleading" and "untruthful", while the payment to the promoter of £25,000 in cash was dubbed "Mower's plunder" and reeked of blatant dishonesty at a time when he was still very heavily in debt to the Banks of Burma and Bengal. Glasgow was particularly suspicious that Rangoon Oil, as the "dividend-earning comet" should link up with the Rangoon Refinery Company, a "bankrupt tail".

Their indignation reached apoplectic heights when yet

another concern, the Anglo–Burma Oil Company Ltd, was registered in London a few weeks later; to Cargill its prospectus was "just about as impertinent and misleading a document as Mower's", while Greenway scornfully bracketed the pair as "those two Burma 'ramps'" in the City of London. The life of Anglo–Burma proved inglorious, thus inflaming the resentment of Cargill and his colleagues that such rackety companies were being encouraged by officials in the name of competition, while reputable and law-abiding ones such as Burmah Oil were subjected to active discrimination.

Rackety or not, British Burmah seemed such a potential threat to the company that Greenway (wrongly) convinced himself and Wallace that Asiatic must have had a hand in its formation. Although he had been managing director of Anglo–Persian since January 1910, Greenway continued for another two years to carry much of the burden of Burmah Oil's external correspondence, and was the company's contact with Cohen. He was elected a London director of Burmah Oil in June 1910, but Cargill would never allow him to join the Glasgow board, refusing to have "any of" him there because of his strong personality coupled with an argumentative disposition. Yet the two men did agree on one thing: that British Burmah would have great difficulty in finding the quantities of crude it needed to keep going without some really rich oil strikes that were quite unforeseeable at that stage. Rangoon Oil's output was actually falling and Cargill believed it would probably exhaust its reserves within a few years.

Burmah Oil still had its crude oil contract with Rangoon Oil but, if British Burmah were to take over the latter's wells, then it might nullify that contract. To forestall such a possibility, Greenway thought the company should perhaps seek an accommodation with British Burmah and offer to market its products in India and Burma: Cargill counselled caution and was backed by Finlay, then on leave in Britain. They need not have troubled themselves. Mower was in no mood to parley with anyone, having received gleeful advice from the geologists about the ample oil reserves in his concessions.

To add to Burmah Oil's travails, the patched-up truce between Asiatic and Standard Oil had broken down, and the two giants were once again locked in a price war. Burmah Oil quickly lost 10 per cent of its market share in India, mostly to

Standard Oil. Annoyingly enough, British Burmah and Indo–Burma gained 3–4 per cent between them, happily making hay during the contest of the giants.

Burmah Oil may have used these reverses as a lever to gain easier terms on kerosene output from Asiatic. Cohen first of all agreed that the company's production limit should be increased by 15 per cent on the 1907 figures. Then, early in 1911, he suggested changing the basis of agreement from fixed quotas to straight percentages of the joint trade. In that way the company would benefit from the steady expansion of the total Indian market, which had grown by nearly three-quarters since 1905/6. The directors pressed hard for a 52 per cent share, but Cohen was set on a 50–50 division.

By April the two sides were still so far apart that Cargill referred the matter to the full Glasgow board, which rejected Cohen's terms. Discussions continued, Wallace using his formidable skills to try to shift Cohen, but even he was able to extract only marginal concessions. The agreement, signed in May, gave no more than a 50–50 share; yet that still benefited the company, as the total kerosene trade in India increased by a further one-fifth to 1913/14 and Burmah Oil's profits more than matched that increase.

That January, the company had recognized the importance of its sales in the subcontinent by appointing for the first time a general manager in India. When the idea was first mooted, Wallace was much opposed to it as a slur on Shaw Wallace & Co.'s hard work over the years. In fact, it was designed to simplify administration, as the main distribution centres such as Calcutta and Bombay had hitherto all communicated independently with Rangoon. For the future, all correspondence would be routed through the general manager in Calcutta, who would also act as a trouble-shooter, able to go anywhere at a moment's notice to sort out difficulties on the spot. Above all, the number of questions referred home would be considerably reduced: a relief at a time when the top directors' work-load had increased all too rapidly following the establishment of Anglo–Persian.

In 1910, Burmah Oil's recurrent worry over crude oil was intensified by news that its own supplies from all sources were barely enough to keep the refineries at full stretch. Yenangyaung remained by far the most important source: in Cargill's words to shareholders, "This marvellously rich oilfield . . . is already unique amongst the oilfields of the world in respect of the extraordinary number of oil sands developed."[3] Yet even there the gas pressure had fallen off so sharply that the wells no longer flowed but had to be pumped. Now that the upper sands had mostly been exhausted, drilling had to be carried down to ever greater depths: below 2,000 feet by 1912, which was nearly twice as deep as in 1905.

An additional problem was that of water seepage. One possible way of shutting out the water would be to cement the wells and then drill them as necessary. That was a familiar device in the Californian fields but attempts to do the same in Burma had failed, partly because the cement would not set quickly enough owing to the warmth of the water. When in 1910 Cargill became seriously alarmed that the brine appearing in a number of wells would contaminate if not ruin the fields, Redwood suggested calling in an expert from California. By July, Thomas Hayes arrived in Burma, strongly recommended for his all-round knowledge of oil matters.

The Hayes episode illustrates two significant problems facing the company at that time. The first was the absence of reliable drilling expertise. Hayes's brief was to inspect the wells and then return to Britain so as to make a personal report. That was intended to help with the second problem, never absent from the directors' minds, of securing an independent assessment of fields work. Several geologists had bitterly criticized conditions there, and the agent was making slow progress in his efforts to impose office control over the fields. The directors made use of Hayes's visit as an opportunity to seek his confidential views on the general conduct of the fields operations. Cargill and Redwood were hopeful that Jacobs, no favourite at home because of what was considered to be his shocking extravagance, could be fired and Hayes appointed in his place. They also hoped to dispense with the fields superintendent at Khodaung, Guy Seiple, who was notorious for his "long-chair attitude" to his duties, having in Rangoon's opinion been "so long in an easy-going groove".

To prepare for these changes Hayes, after spending several months in Burma and then discussing his findings with the directors in Britain, was sent back to California to recruit drillers and purchase new equipment. He was then to return to Burma and begin putting the fields into shape. When the first contingent of new drillers arrived in Burma, Guy Seiple was given notice and told to be out of the fields within 12 hours. This precipitated a walk-out and the threat of a strike by the whole of the fields staff; but he went quietly and all were soon back at work.

Finlay, having been on leave during his previous visit, soon became suspicious at the contrast between Hayes's very impressive talk and his lack of practical capability, being unable even to cement wells properly. After a very candid interview he unmasked the man as a fraud and cabled the directors at home to recall him. This they did: Redwood took full responsibility and agreed to give Hayes his *congé*. Yet Redwood at 65 was not the man he had been in his prime, and Cargill doubted if he would be as abrasive as circumstances demanded. Later revelations about Hayes's bragging and untruthfulness whilst in Burma provoked Finlay into an intemperate expression unique in his correspondence: "a most dangerous swine". One man did not conceal his rejoicing at the turn of events. "Jacobs' tail is a little too far up over Hayes' dismissal," reported Rangoon. Jacobs stayed as fields manager for another five years.

The whole episode, humiliating as it was for the company, only highlighted the lack of adequate supervision in the fields without doing anything to resolve it. Drilling costs in 1910 were over double those in 1907, and the cause was not merely the greater depths or the need to pump, but also sheer extravagance. Tools and equipment were often scrapped while still perfectly serviceable, or left lying about to rust. Lubricating oil for the steam engines was wasted by being poured out from old jam tins, so that the introduction of proper oil cans markedly reduced consumption. The manager of the fields stores department was unable to control wastage because he did not have sufficient technical expertise to know how much material the drillers really needed.

Geological effort, too, left much to be desired. A very much overdue reform was carried out when Sir Thomas Holland, lately retired from the Geological Survey of India, became joint geological adviser, with Cunningham Craig, to Burmah Oil. In 1912 he visited Burma and helped to reorganize the geological office at Nyaunghla, establishing an adequate filing system and

proper well records. The filing system would ensure that all
relevant reports were immediately available for consultation.
The availability of drilling journals, too, would allow the head
geologist to recommend to the fields agent, now responsible for
broad drilling policy, any change of plans that the journals
might indicate.

At that time, the company was making intensive geological
probes throughout the province, described as "the last big
'sweeping' operations" in an almost desperate search for more
crude. By mid-1911 the company was reckoned to have fairly
complete knowledge of about 15,000 out of the province's
250,000 square miles. This led Cunningham Craig, the shrewd-
est if scarcely the best-loved of the company's technical staff, to
conclude that there was "uncommonly little chance of finding
anything in the shape of a new paying field in any accessible
district" in Burma. The only crumb of comfort he could offer
was that "our opponents have an infinitely smaller chance of
doing so."

Wallace had been following all these efforts with sceptical
interest. As early as September 1909, when Finlay had first
proposed from Rangoon that the company should mount the
country-wide geological sweeping operations, he wondered
aloud if that was a rational policy or one put forward out of pure
habit. With his usual inexorable logic he linked this issue with
the local governments' policies of discriminating against the
company. Money spent on prospecting was, he affirmed,
"worse than money merely wasted"; for it told competitors
whether oil was to be found in any field or not. As long as
governments insisted that competitors should have the right to
move into any new field that might be discovered, there was
little inducement to energetic exploration activity. But if the
known fields ran out before new ones could be found, then all
the rival producers in Burma would be "killed" before Burmah
Oil:

> We should have reserves of oil in Persia which the others have not
> got at all, some cash to tide us (with our Persian supplies) over the
> bad time in Burma – the possible bad time between the giving out of
> the at present known fields and the finding of the hypothetical new
> fields.

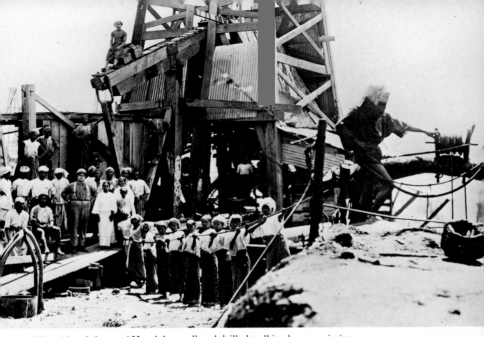

17. The old and the new! Hand dug well and drilled well in close proximity.

18. Team of Burmese well diggers, 1910. Note helmet made out of kerosene tin, introduced c. 1900.

19. Hand dug well at Yenangyaung with the driller about to be lowered by a team made up of members of his family. Note the western-style umbrellas.

20. Clearing out oil bore at Digboi, 1905.

21. Oil tank ablaze, Syriam
refinery, 1909. Note pot-still in
foreground.

22. Sir Campbell Kirkman Finlay
(1875–1937), director 1914–1937.

23. Finlay Fleming & Company assistants, 1911. C. K. Finlay is seated, centre, with, on his left reading left to right, Duncan Garrow, R. I. Watson and Captain G. W. Currie. A. B. Ritchie is standing top row left.

24. William Knox D'Arcy (1849–1917), director 1909–1917.   (A BP photograph)

25. Robert Irving Watson (1878–1948), managing director 1920, chairman 1943–1947.

Although his arguments did not alter his colleagues' determination to continue prospecting, he returned to the attack in March 1910, after the government of India had raised the tariff on kerosene by half as much again, totalling about 20 per cent of the price: the first change that had been made since 1894. Now that, in his opinion, the tariff was high enough to cut out foreign competition, he proposed that Burmah Oil should at once come to terms with all rivals in Burma and rescind the agreements with Asiatic forthwith. "While the other Burma companies have free hands," he maintained, "the Burmah Oil Company is paying a subsidy to the Asiatic Petroleum Company for absolutely no *quid pro quo*." Cargill rejected that proposal; for all his conventional cast of mind, his judgment was sound here, as the enmity of Asiatic was far more to be feared than that of rival companies in Burma.

A more pressing worry for the company was that the tariff increase might be followed by the imposition of an excise duty on kerosene produced in the Indian empire. The India Office still believed that the indigenous companies' tariff advantage should be offset by an excise duty as soon as those companies were strong enough to stand on their own feet, and Burmah Oil was by then doing well despite its financial losses in providing fuel oil for the Royal Navy. That time had probably arrived and in August, D'Arcy – who was then for some reason the Burmah Oil director most highly regarded by the government of India – wrote to ask the viceroy-designate, Sir Charles Hardinge, if he would receive a deputation before leaving for India to hear the company's objections to an excise duty. Hardinge refused and insisted that all representations must go through the usual channels. That question was to flare up again in 1911–12 (see Chapter XIV).

In the summer of 1911 British Burmah was reported to be in deep trouble. It had made a large operating loss during its initial year, while its subsidiary Rangoon Oil, which before the merger had declared dividends of 50–70 per cent, was passing its dividend altogether. The Burmah Oil directors talked big among themselves about the kind of assistance they would offer if and when the collapse came. The favourite suggestion was a secured loan to Rangoon Oil, representing a cash advance against future deliveries of crude; that would serve to keep alive the current agreement with Burmah Oil.

Cohen, normally so anxious to swap information in periods of stress, was for once uncommunicative and "very much engaged", hopping over to see the Royal Dutch people just when Shell shares were booming on the London stock exchange; thus he almost certainly had "some *coup en train*". Whatever it was, he kept mum. It could scarcely be an accommodation with Standard Oil, which gave no sign of any disposition for peace at a time when it was seriously injuring the earnings of Burmah Oil and Asiatic in India.

Then in November, Finlay cabled the dramatic news that the Bank of Burma had closed its doors. Mower and Clifford were the bank's most prominent directors, and never in the history of commerce could the downfall of any such institution have been so universally and accurately forecast. Finlay and other local correspondents had in recent months almost run out of suitable epithets to describe its deteriorating condition. "Shaky", "nervous" and "very sick" were a few of them. Clifford had telephoned Finlay that September to ask if there was any truth in the rumour that France and Germany were at war. As Finlay reported home, "He remarked that he hoped to heaven it is true and that a German warship would pay a visit to Rangoon and help them to settle up their difficulties!"

No gunboat put in an obliging appearance, and the actual collapse occurred through an act of unbelievable folly, or worse. Rangoon Oil had already exhausted its credit with its main bankers, the Bank of Bengal; and so Clifford went to another bank for help, only to be called a "damned liar" by the manager during a stormy interview. Mower and Clifford then turned to their own Bank of Burma and borrowed the equivalent of £66,000 without informing the bank's trustees in London, which as directors they were legally bound to do. Only a day or so before the crash they converted the temporary overdraft into a four-year term loan. The network of deals had been so far outside the law that the pair could be liable to serious criminal charges.

Cargill and his colleagues indulged in some alluring pipe-dreams about how the bank's ruin would inevitably bring down Rangoon Oil and with it British Burmah. With the full blessing of the local governments, they themselves could then take over those companies' assets and activities: that course would have every advantage over earlier schemes to hobble Rangoon Oil through buying up its shares. The chairman of British Burmah had come out post-haste to Rangoon when his company had been reported to be "very dicky". He had at once called on Finlay and frankly admitted his grave predicament, and asked for financial help. The bank's demise had followed very shortly afterwards. Cargill judged that if one or both of the local banks had to foreclose on Rangoon Oil in an effort to save themselves, Burmah Oil was about the only organization that could bail the victims out.

In the event, all that the company gained from the débâcle was a nasty shock. Alarmed for their subsidiary's future, the British Burmah directors hurriedly sought to transfer into their company's name the wells held by Rangoon Oil. The local government did not even demur, despite a vigorous protest lodged by Finlay. Officials' replies were so evasive that he could only regard such a complacent attitude as evidence that they did not intend to give Burmah Oil fair treatment. The company therefore considered bringing a civil action against Rangoon Oil for breach of contract, to enforce deliveries of crude oil. The legal issues were particularly complex; yet, declared Finlay, "I am all for fighting them, even if our case does not look pretty."

No civil action took place and, as 1911 drew to a close, Burmah Oil's main preoccupation was that its earnings from Rangoon would be down by a quarter of what they were in 1910: they were in fact the lowest since 1905, before the original kerosene agreement had been signed. Yet the news from the east, appropriately conspiratorial as it was, gave a hint that the current price war might at last be ending. Charles F. Meyer of Standard Oil was reported to be on his way to Calcutta and Greenway inferred that some "important development" was being hatched.

How Meyer or his superiors restored peace to the much battered kerosene market in India is not recorded in Burmah Oil's archives; perhaps Cohen explained it verbally to the directors, with little being committed to paper. Undoubtedly it was merely one aspect of a world-wide accommodation. Although a few minor scares occurred, there were no more price wars until very different economic conditions prevailed in 1927–8.

Notes

1 For the peculiar characteristics of oil, which in his view necessitates combination rather than outright competition see P. H. Frankel, *Essentials of Petroleum: A Key to Oil Economics* (2nd ed. 1969), especially Part V, ch. 2. cf. Sir R. Waley Cohen, "The Economics of the Oil Industry", *Journal of Institution of Petroleum Technologists*, X, 1924, pp. 371–90.

2 Quoted in A. Sampson, *The Seven Sisters: The Great Oil Companies and the World They Made* (1975), p. 58.

3 AGM 1914, *Glasgow Herald*, 3 June 1914.

CHAPTER XII

Persia
1909–14

In the early months of 1909 the Burmah Oil directors should have been more than satisfied with prospects in Persia. The oilfield had been shown to be both extensive, up to 10 square miles in area, and prolific, at a time when the long-term future of oil supplies in Burma was already seriously in doubt. Anglo–Persian's concession was not only unique in its magnitude but also mercifully free from government interference and the consequent need to lobby uninterested or hostile officials. There were no rival companies in adjacent blocks to eye suspiciously, no fire hazards caused by wells sited too close to one another – for Redwood had laid down that Persian wells should be at least 500 feet apart – with no need for an offsetting policy. Yet despite the set fair climate, Cargill for one was to find the next few years perhaps the most anxious in the whole of his care-laden life.

Although the crude oil was there in profusion, it had little value until it could be transported to the coast, refined and then delivered to appropriate markets. To help solve these and many other questions, an effective local organization was needed in Persia. Communications problems ruled out a branch office directly responsible to London and it was clear that a firm of managing agents was needed. The mercantile houses on the spot, such as Lynch Bros, were found wanting and a new one was therefore set up from scratch. The first two men appointed, (Sir) J. B. Lloyd and C. A. Walpole, were from Shaw Wallace & Co., and the agency was named Lloyd Scott & Co. (the latter name apparently commemorating its Scottish origins).

Reynolds, the main loser from these changes, was not told of them until mid-May when he was about to go on leave. As an intensely proud man he must have felt mortified at being abruptly downgraded to the position of general fields manager. He replied to the news in rather surly fashion, pointing out that his own headquarters were at Ahwaz and that no suitable accommodation for Europeans existed in the mud-brick houses of Mohammerah, where the agency was to be based.

Many changes took place while Reynolds was on leave. When he returned in December 1909 he found that the refinery and pipeline were being actively planned without any advice from him; worse still, he was to become the slave of paperwork now that the agency was well and truly established. He was soon at loggerheads with Lloyd Scott & Co. and in particular with the ex-Shaw Wallace man who was manager of the Mohammerah office. In his new and restricted role, he decided to do as he chose. He refused, for instance, to act any longer as political agent with the Bakhtiari tribes, and he paid the directors back for their earlier secretiveness by keeping them in the dark about what was happening on his patch. Now that relations between him and Mohammerah were so bad, those at home had little hope of securing reliable information via that office.

By October 1910 the Anglo–Persian board had to authorize Greenway, now managing director in Wallace's place, to demand assurances from Reynolds that drilling plans and other fields work would be brought up to date well before the pipeline and refinery were due for completion. Moreover, he must submit progress reports on his various operations, together with estimated dates of completion. Yet even before the cabled reply reached him, Greenway was convinced that he and Hamilton ought to go out together, so as to see for themselves what was really happening and also to reinforce the local organization.

Greenway's determination was strengthened when Reynolds's cables arrived, giving few hard facts. He at once recalled Reynolds, who left Mohammerah in January 1911 shortly before the visitors arrived there. In one of his earlier brushes Reynolds had written, "When we have success, nothing will give me greater pleasure than to learn that some of the directors of Messrs the Concessions Syndicate Ltd. have been there to judge for themselves"; he was now to be denied the exquisite pleasure of receiving personally a delegation from Messrs the Anglo–Persian Oil Company.

They did not prejudge the issue, but what they discovered could only confirm their worst fears. As they later reported to their colleagues, Reynolds had done little to develop the fields, and had disregarded earlier instructions to start erecting staff accommodation and carry out other essential works. After talking to a number of people, they reluctantly concluded that he had been deliberately putting a brake on operations so as to be able to claim that the managing agents and others outside his control were entirely responsible for the mess, and that the only remedy would be to give him back his old job of supremo. After all these years it is impossible to establish the precise truth, but Reynolds would certainly have been quite capable of thinking out such a scheme and biding his time until it came to fruition: the dark side, so to speak, of his matchless patience and resourcefulness over the preceding six or seven weary years.

While recognizing the "magnificent work" he had earlier done, which rightly entitled him to the "very highest praise", they decided Reynolds would have to go. When he arrived home, he was interviewed by Wallace, who no doubt put on an appropriately crusty manner; in February, Reynolds's engagement was terminated and he was free to resume his pioneering activities elsewhere in the world. He lived until 1925, occasionally featuring in the Burmah Oil correspondence as he turned up in various parts of the world as adviser to rival firms. His remarkable role as a pioneer in Persia is unquestioned: no amount of ink spilt since 1908 for or against him can diminish that high achievement.

By the early months of 1911, the 140-mile long pipeline was nearing completion, directed by Charles Ritchie, who had also built the one in Burma (see Chapter VIII). It was ready in mid-year. Greenway and Hamilton decided to bring it under the fields management, where Ritchie was placed in overall charge. This new post was similar to that introduced six years previously in Burma, the only different circumstance being that Ritchie was an engineer and not an office man. For the moment only small quantities of crude were pumped through the pipeline, for one delay and difficulty after another dogged the construction of the refinery.

Andrew Campbell was its designer, assisted by John Gillespie on the engineering side. He recommended the Henderson-type stills, as improved by himself, which had worked so reliably in Rangoon. The first consignment of plant had not even arrived in Abadan until the summer of 1910 and, as there was no skilled labour within miles of the site, Glasgow had to arrange with Finlay Fleming & Co. for trained artisans such as fitters, masons and riveters, as well as clerks, to be despatched from Burma. They were mainly Indians from the Rangoon refineries. Not until July 1912 could the refinery be given a trial run-through, whereupon various sections immediately broke down. Once again the producing wells had to be throttled back to curtail the flow of oil, and for the rest of the year the refinery's throughput reached only half its capacity.

In this depressing situation, with crude still seeping from the throttled wells and having to be burnt once the storage tanks were full, the directors were thrown back on the most humiliating expedient of all: they had to seek marketing agreements with Asiatic. Greenway was the negotiator, in consultation with Cargill. He approached Cohen as soon as the pipeline was completed, and made a preliminary agreement over crude oil in April 1912. That was followed by a more comprehensive one in October to cover kerosene and petrol as well.

The terms agreed over prices and quantities were familiar to Greenway from his days of helping out in Burmah Oil, but the main difficulties were caused by fuel oil. Despite Cohen's strong objections, Greenway secured complete freedom to negotiate independently with the Admiralty and also with the Indian railway authorities, whose locomotives could burn either coal or oil. An agreement on fuel oil, although for a very modest amount, was finally reached with Asiatic in November 1913. These arrangements were to last until the end of 1922; Asiatic, but not Anglo–Persian, could terminate them earlier. Greenway later claimed to the Admiralty that he had signed under duress; but Asiatic could well have driven an even harder bargain than it did, as Anglo–Persian's negotiating position was, to say the least, far from strong.

Indeed, Anglo–Persian's difficulties were soon to land it in an alarming financial crisis.[1] During the first three years up to March 1912 its annual expenditure averaged £250,000 a year, of which three-quarters went on capital items such as constructing the refinery and pipeline, and on fields development. That was not so different from the level of Burmah Oil's spending at the time, but Burmah Oil was remitting home a good profit of £500,000–£750,000 a year. By contrast, Anglo–Persian earned nothing, let alone the gross profit of £270,000 a year forecast in the prospectus, and therefore had to find the necessary cash from the preference share and debenture issues of 1909. Towards the end of 1911 all the net proceeds, totalling £800,000, had been exhausted, so that an extra £300,000 worth of preference shares had to be issued: perhaps enough to see the company through another year, but in any case it involved a further cash drain of £18,000 a year for Burmah Oil, which had to guarantee payment of all preference share interest.

Cargill became increasingly despondent over Anglo–Persian's recurrent technical problems and the financial troubles that ensued. In February 1912 he lamented to Finlay, "This Persian business seems to get more complicated every day, and I am much exercised over the position generally." As if that were not enough, his fellow Anglo–Persian directors seemed hell-bent on dragging that company into a new and potentially even more ruinous "adventure" in Turkish Mesopotamia. On 28 August he wrote a very stern letter to Greenway on the state of the Anglo–Persian finances. It is quoted at length in Chapter XIII, but the blunt message was that Burmah Oil could not be expected to bail Anglo–Persian out indefinitely.

That letter signalled a turning-point in Anglo–Persian's history. Now that its surplus crude and all products except naval fuel were to be sold through Asiatic at not particularly remunerative prices, Greenway was pinning his greatest hopes on that fuel oil trade, as Cargill mentioned critically in that same letter. Yet it will be remembered that as early as 1908 Wallace had affirmed that if the navy needed fuel oil before the "big company" was earning adequate profits, the Admiralty would have to remunerate the company fully for all the trouble and expense involved. That was the last thing the Director of Navy Contracts, happily buying cargoes of oil here and there in the cheapest markets, wished to do.

To be sure, there were indications that the Admiralty might well be forced soon to change its purchasing arrangements. Ever since the Oil Fuel Committee had been disbanded in 1906, it had no standing advisory committee of petroleum experts, and Sir John (now Lord) Fisher, although First Sea Lord until 1910, had done his cause of championing oil no good by a clamorous advocacy of the internal combustion engine, which at that time was purely experimental as a possible method of powering large vessels. Yet soaring quantities of oil were to be needed not only for destroyers and submarines but also for major warships if they were to realize the potential advantages of speed and overall efficiency.

Now the apprehensions of a business man (Cargill) and the strongmindedness of a statesman were to bring together the oil policies of both Anglo–Persian and the Admiralty, to the ultimate advantage of both. In October 1911 the indefatigable Winston Churchill became First Lord, and one of the earliest questions he raised concerned the state of the navy's oil supplies. The unsatisfactory reply, after six years of Fisher, led him to set up a departmental committee under the Fourth Sea Lord, Captain (Sir) William Pakenham, who was responsible for supplies and transport. Its primary function was to advise on how to ensure a large enough supply of fuel oil so that all future warships could be constructed to burn oil.[2]

The Pakenham committee interviewed a number of oil magnates, including Strathcona and Greenway as well as Deterding and Samuel. Strathcona contented himself with some rambling reminiscences, leaving it to Greenway to reveal the enormous oil possibilities in Persia. Offering to negotiate a contract of 50–75,000 tons (313–470,000 barrels) of fuel oil a year, Greenway gave no hint of the alarming technical setbacks thus far encountered at the Abadan refinery.

He also stressed the financial aspect, broadly along the lines of Wallace's observations of 1908. As the quantity of fuel oil taken up from Rangoon was less than three-quarters of the budgeted 10,000 tons a year, the total cost of that fuel oil was £4.00 a ton, whereas the Admiralty was buying it at £1.00–£1.25. The committee saw that there was some justice to his case, but instead of making recommendations, concluded that the wider question of oil supplies should be referred to a royal commission.

Winston Churchill duly set up the Royal Commission on Fuel and Engines in July 1912, Fisher being an obvious if not ideal choice as chairman.[3] The members included George Lambert, Civil Lord of the Admiralty, and Admiralty naval and civilian officials, as well as Sir Boverton Redwood and Sir Thomas Holland; those two last names calmed Cargill's fears that the case for financial assistance to Anglo–Persian might be lost through an unsympathetic hearing.

Yet royal commissions take their time in gathering evidence, and meanwhile Greenway felt he must at once canvass official-dom after Cargill's clear warning. As Chapter XIII shows, Greenway was already busy conferring with government de-partments over Mesopotamia; then very shortly after receiving Cargill's letter he disclosed to the Foreign Office that Shell had for some years been striving to absorb Anglo–Persian.[4]

To be sure, Cohen had several times lately made tempting offers to buy his way in. When towards the end of September Greenway met Admiralty officials, he warned that Anglo–Persian would probably lose its independence unless the government could help to subsidize its expansion in order to ensure the required quantity of naval fuel. His perhaps over-dramatic statements he later justified by asserting that, although his board contained men such as Strathcona and D'Arcy who would resist a take-over "on imperial grounds", others – notably Cargill – thought only of shareholders' in-terests and might therefore be inclined to cave in to Shell.[5]

To officials Greenway portrayed himself as the one who relied on the government to rescue Anglo–Persian from Shell's clutches by granting adequate financial assistance. He pro-posed that the Admiralty should contract to pay £100,000 a year over twenty years for fuel oil, which it would be able to buy on advantageous terms. Such a firm contract would permit Anglo–Persian to raise £2,000,000 on the stock exchange, something it was quite unable to do on its own as long as it was earning no profit. Should parliament veto it, then the govern-ment of India might be persuaded to provide a similar long-term contract to purchase fuel oil for its railways, and offer some to the Admiralty.

On learning of these proposals, the government departments concerned split along predictable lines. The permanent under-secretary at the Foreign Office, Sir Arthur Nicolson, took the realistic view, prompted by continuing concern about Russian designs on that region, that if no alternative solution proved feasible then the government would have to pay up. The Admiralty, being a large spending department and therefore subject to Treasury discipline, refused to consider the terms. Nicolson therefore requested his minister, Sir Edward Grey, to take the whole question to the cabinet or Committee of Imperial Defence as soon as possible, since it was clearly of "such supreme importance". Grey insisted that, in accordance with normal government practice, it must first be discussed between departments at official level.

As Cargill was leaving for the east towards the end of November, the whole of that month was exceptionally busy. Anglo–Persian's funds were so depleted that the Burmah Oil board authorized a temporary loan of £100,000, Strathcona and D'Arcy between them contributing an extra £50,000. Cargill gave evidence before the royal commission, speaking mostly about Burmese supplies and stressing that prospects in the fields there were so poor that really large quantities of naval fuel oil were out of the question.

On 8 November, with pressure of work building up daily, Cargill was rattled by the news that Cohen had written to Finlay, now home from Rangoon and recently appointed a director of Anglo–Persian, demanding Greenway's dismissal from the managing directorship. Cohen accused him of "going about describing the Asiatic group in an exceedingly hostile way". That was a clever attempt to split the various interests in the Anglo–Persian board, and it was true that the incorrigible Greenway had been asking for trouble over the years with his anti-Shell outbursts. He should be replaced, said Cohen, by a man prepared to carry out "with loyalty and goodwill the policy of those directors of the Anglo–Persian Oil Company who represent the Burmah Oil Company, to whom the Anglo–Persian Oil Company of course owes everything."

Yet when Cohen travelled to Glasgow to put pressure on Cargill, he found the latter unwilling to budge an inch. "I had two hours' very straight talk with Cohen this afternoon and I think I have succeeded in putting matters generally on a more satisfactory basis," the relieved Cargill reported to Finlay the same day. No doubt on Deterding's instructions, Cohen resumed the offensive shortly before Cargill's departure when they met in London to clear up outstanding points on the fuel oil agreement. He hinted that if Asiatic were barred from forging closer links with Anglo–Persian, the Shell group would use its market strength to start a price war in the east that would quickly bankrupt Anglo–Persian.

If his intention was to give Cargill plenty of food for thought during a prolonged absence from Britain, then he succeeded all too well. The month-long sea voyage did nothing to dissipate the chairman's worries; they in fact intensified and cast a shadow over his normally agreeable sojourn out east. Late in January 1913 he wrote home to Finlay from Rangoon and bared his soul in a way that is almost too painful to read:

> I can't get my mind away from business here and neither feel fit nor in the mood for late nights, especially with people I never saw before, don't give a damn for and will probably never see again! You can imagine . . . what my feelings are as I read in your home correspondence and [Andrew] Campbell's letters what a hell of a mess Persian things are in. It's all very well to say "don't worry," but my name and business reputation are too closely associated with the Anglo–Persian Oil Company to admit of my not being terribly anxious and worried over the present horrible state of affairs.

> It's a great great pity of course that I am constituted like this, and as I have said a hundred times I ought never to have been in the position I am in. But it seemed impossible for me to keep out of it, and the pity is people don't understand my nature better and make proper allowances for it.

Letting off steam in that way clearly did him some good. By the next mail he apologized for his habit of viewing things through "black spectacles" and claimed that he now felt in "a better and more reasonable frame of mind". Yet his timorous attitude sharply contrasted with the mounting self-confidence of Greenway who was energetically pleading his cause up and down Whitehall. There only one department had initially been at all responsive. After an inconclusive interdepartmental meeting in November, the Foreign Office had written to the Admiralty, arguing that financial as well as diplomatic backing was essential to preserve Anglo–Persian's independence. The Admiralty was not to be moved, and in January the India Office, also appealed to, refused its support on the ground that no specifically Indian interests were concerned. The Foreign Office could only express to Greenway the pious hope that the company's interests in Persia would not be surrendered to foreigners, adding more earthily that if they were, then any government backing would be at an end. Greenway argued back, but the end of that particular road seemed to have been reached.

Even at that moment, however, Admiralty officials were being forced to reconsider their hard line towards Greenway's proposals. Stirred up by pressure from Churchill for estimates of future oil requirements in relation to known reserves,[6] and concerned about rising prices now that world demand for oil products was on the increase, they modified their policy of buying fuel oil on an *ad hoc* basis and in February for the first time publicly sought tenders for forward contracts. One of the earliest responses was from Anglo–Persian, which offered 30,000 tons between September 1913 and June 1914. Then Deterding, in accordance with an undertaking he had made to the Pakenham committee, signed a two-year contract for substantial quantities over the period 1914 to 1916. Even so, the navy's urgent need for steadily increasing oil supplies made Greenway optimistic that officials would very soon have to open serious talks with him.

Sure enough, by the end of February the Admiralty was asking him what security he could offer against any advance payments it might be willing to make for future oil deliveries. On 6 March he replied that the best security would be a specific charge on its assets. During that week he had three separate discussions with Sir Frederick Black, the Director of Navy Contracts, and brought along Sir Frank Crisp as his company's solicitor to advise on suitable collateral. Then on 13 March Winston Churchill, together with Sir Francis Hopwood, a civil lord, and Black, met Greenway, Finlay also attending in Cargill's absence overseas. As this was apparently the first time Churchill had met any of the directors concerned, he may just have wished to look them over for himself. At that stage he raised only very general points, stressing that his mind was still open.[7]

As he had earlier anticipated to Finlay, Cargill found on arriving home more than enough to dismay him. The £150,000 advance by Burmah Oil and the others was by then almost exhausted, and a board meeting held in Glasgow shortly afterwards had to authorize the lending of "further sums as required up to a reasonable amount", the word "reasonable" being undefined. But what really shocked him was his discovery that even cautious directors such as Hamilton were "quite prepared to go the whole hog with Greenway" in confidently offering substantial amounts of fuel oil – up to 500,000 tons a year – at a time when Abadan was producing few products that were saleable at all.

Yet Greenway, having already covered so much ground with the Admiralty, was certainly not willing to back down. Just as in 1905 the Burmah Oil directors had taken pains to instruct Admiralty officials in the realities of commercial life – in that case, such vital matters as storage costs – so now Greenway sought to convince them that as Anglo–Persian would be earning all too little from the product agreements with Asiatic, the terms of any fuel oil contract with the Admiralty must be reasonably generous. Meanwhile everything depended on what Churchill had in mind. Three months passed before he took the plunge and submitted a cabinet paper in June on "Oil Fuel Supply for His Majesty's Navy".[8]

To build up reserves and safeguard future flows of oil, he wrote, a series of forward contracts would be essential. Ideally, those should come from sources under firm British control or influence and be remunerative enough to maintain independent competitive firms in existence, thereby forestalling the risk of a world-wide oil monopoly. An accompanying Admiralty memorandum stated that if the government decided to pursue the question of a guarantee, then it might have to impose on the supplier provisions for profit sharing and some control over management decisions.

On 9 July the paper was brought before the cabinet, which agreed in principle that the government should "acquire a controlling interest in trustworthy sources of supply, both at home and abroad". It also set up a cabinet committee to make detailed recommendations. Within a fortnight Black informed Greenway of its conclusions: that government would guarantee Anglo–Persian the whole £2,000,000 or alternatively take up some of its capital. As either proposal required legislation, which could not pass through all stages before mid-1914 at the earliest, Burmah Oil would be expected to bridge the financial gap by guaranteeing a fresh issue of preference shares or by providing cash. So Anglo–Persian need lose no time in ordering new equipment through lack of ready money.

The cabinet committee also proposed that a delegation should be sent to Persia and study at first hand the prospects for future output "in order that every step possible had been taken to assure themselves of the Anglo–Persian Oil Company's ability to implement the contract." That was Greenway's interpretation of the proposal, as reported to Cargill. Admiral Sir Edmond Slade was chosen to lead the delegation, which also included John Cadman, mining professor at Birmingham University, and E. H. Pascoe, the oil expert with the Geological Survey of India.

Slade's qualifications for this assignment are not at all clear. He had earlier served as Director of Naval Intelligence, only one rung below the Sea Lords, until Fisher had "got rid of" him as a "fool" and banished him to command the East Indies fleet.[9] He had somehow come back into the Admiralty for "miscellaneous service". So anxious was Churchill to push matters to a rapid conclusion that he expected the delegation to leave without delay, but he had to be told that they would merely be arriving in the midst of a torrid Persian summer when little could be seen or done. He therefore reluctantly agreed to their departure being postponed until October.

In order to discuss some preliminary points, at the end of July a top-level meeting was held at the House of Commons. There most of the executive directors of Anglo–Persian, namely Cargill, Wallace, Greenway and Finlay, met a formidable establishment team comprising Churchill and Hopwood, Lloyd George (Chancellor of the Exchequer), Sir John Bradbury (Secretary of the Treasury) and Walter Cunliffe (Governor of the Bank of England). Cargill, as Anglo–Persian's ultimate paymaster, was not at all overawed by this august gathering; nor did he show much liking for the proposals set before him. Those were that the government should make a £2,000,000 advance with the option of converting the advance, fully or in part, into debentures or alternatively enough shares to give it overall control. Should it choose not to exercise that option, then Burmah Oil would itself have to guarantee both interest and repayment of the advance.

Cargill would have none of those conditions. As it had been his ultimatum to Greenway that precipitated the negotiations, he could hardly be expected to saddle Burmah Oil with further heavy liabilities on the fiat of government. He therefore refused to contemplate the prospect, however improbable, of accepting a £2,000,000 commitment on top of the £1,500,000 that Burmah Oil had already sunk in Persia. The government team trotted out the familiar argument that they could scarcely be expected to shoulder such risks if the company had less than full confidence in the Persian operations. With the two sides so far apart, negotiations broke down.

Later historians have tended to regard the Slade delegation as simply a public relations exercise. Yet the Burmah Oil directors did not see it that way at the time. Wallace believed that, whatever might be said by officials, the delegates had almost certainly been instructed to report in confidence on what they happened to see, and therefore might have some harsh things to say about the poor state of the refinery. In mid-August, a fortnight after negotiations had become deadlocked, Wallace, Greenway and Finlay sat down together to decide what should be done to put Abadan in order.

Finlay, well informed by his regular correspondence with Persia from Rangoon and more recently from the London office, condemned the Abadan refinery outright as a "scrap heap". Wallace rather mischievously repeated that "picturesque phrase" to Glasgow and greatly incensed Cargill, who demanded of Finlay the reasons for making what seemed to be "a most extraordinary and incomprehensible statement". Finlay with his usual deftness took the heat out of things by explaining that he was referring to the men involved rather than to the plant: "If the refinery is not properly organized and run," he explained, "it may be considered a scrap heap as far as results are concerned."

The refinery was plainly in a deplorable condition at that time. It should have been ready two years before, but had not yet run at full capacity. Of the products it did turn out, the petrol was not marketable, the kerosene was too smoky and much criticized by customers; to crown it all, the naval fuel oil failed to pass the Admiralty test, because too many of the lighter fractions were being taken out and the flashpoint was too high. There had been "one chapter of misfortunes after another" since its opening, mainly because everything was done slackly and without foresight, with "no cohesion and no go" among the staff. The works manager himself was weak, and the top man at Mohammerah, equivalent to the general manager in Rangoon, had not the qualities to oversee the refinery.

The directors therefore appointed a refinery overseer: Andrew Gillespie, formerly a fields agent in Burma who had won a reputation for tact and firmness. He and a new works manager arrived a little ahead of the Slade delegation, as Wallace wanted delegation members to see for themselves that new brooms were already on the sweep in Persia. Gillespie's initial estimate was that it would take six months to do all that was needed. Yet Cargill did not complain about the further delay, even though it came on top of the stalemate in the Admiralty negotiations. Twice more he persuaded the Burmah Oil board to authorize extra advances to Anglo–Persian, finally up to the maximum allowed under the latter's Articles of Association.

Greenway from time to time prodded the Admiralty, but was repeatedly told that nothing could happen until Slade had reported on his mission after its return in January 1914. That report turned out to be entirely favourable and even enthusiastic, the concession being hailed as "a most valuable one" which if "judiciously worked would probably safeguard the fuel supply of His Majesty's Navy". The government was therefore recommended to provide the necessary capital. Barely a fortnight after the mission's return, therefore, Churchill obtained cabinet permission to open negotiations with Burmah Oil in order to secure a majority shareholding in Anglo–Persian.

The delegation even spoke well of Finlay's "scrap heap", commending the refinery as "of modern construction, well designed, and laid out with a view to future extensions".[10] Even so, various troubles afflicted the Persian operations during the early months of 1914. Yet Gillespie's patient but resolute overhaul of refinery organization gradually bore fruit, throughput at Abadan in the first quarter of 1914 being almost three times that for the corresponding period of 1913. Many of the technical difficulties, too, were being overcome, with marked improvements in the quality of saleable products.

In January 1914 Lord Strathcona died, aged 93. He had been a conscientious and genial figurehead, there when needed and lying low when he was better out of the way. Above all, he had played his unobtrusive part in making the company acceptable to Whitehall. Cargill and Wallace, as the senior directors, therefore sought to preserve good relations with government circles by tactfully sounding out official views on a successor.

One of the first names put forward, but discarded, was that of Strathcona's cousin and virtually a carbon copy: Lord Mount Stephen, who had made his fortune in Canada and at 85 was appropriately full of years. Lord Scarbrough was then asked, for no apparent reason except that he was a tenth earl, and after him Austen Chamberlain and Lord Selborne. The last two were active politicians and had no wish to give up their parliamentary work. By then it was the end of June and the post had been vacant for five months.

Wallace therefore sought guidance from Black at the Admiralty who, after consulting Churchill, proposed that Wallace himself should take the job. He declined on health grounds, and was in fact planning to get away shortly on a prolonged rest-cure. Black then suggested Greenway, since he possessed "special knowledge of the company's affairs, as well as of the local situation in Persia and of all the aims and details of the government intentions". Their Lordships had been very impressed with his able and vigorous conduct of recent negotiations and were sure he would run the company on successful commercial lines and also safeguard the government's specific interests.

Wallace at once replied giving his cordial assent, and only then informed Cargill, writing shrewdly in his usual semi-ironic manner:

Had they [the authorities] named a "personage" we might have struck one who would have been content not to hinder work by unduly interfering and possibly seeking to impose impracticable ideas of business and procedure or distasteful ones to us. Or we might not have been so fortunate. In either case the work would not have proceeded more smoothly nor yet more in accordance with our views than it will do with Charles Greenway in the chair.

He then virtually dared Cargill to veto the proposal. Cargill raised no objection and, as there was no obvious candidate for the managing directorship, Greenway was allowed to hold both posts together.[11]

In the final stages of negotiations over Anglo–Persian, two points preoccupied the Burmah Oil directors. How precisely would the government relieve them of their massive financial burdens, and what control did it, as majority shareholder, propose to exercise over Anglo–Persian's affairs?

The first point was covered by the agreement with the Treasury and the Admiralty, signed by Cargill, Wallace and Greenway on 20 May 1914.[12] The government was to take up 2,000,000 newly created £1 ordinary shares, as well as £1,000 nominal of preference shares and £199,000 nominal of debenture stock. The extra £200,000 arose from an estimate by Anglo–Persian of the cost of additional refinery capacity, a new 10-inch pipeline, geological surveys and fields development, and shares in a new tanker company. On the same day a fuel oil contract, to run for twenty years, was signed with the Admiralty. Perhaps in response to Cargill's remonstrations, the quantities were more modest than Greenway had earlier suggested: 50–70,000 tons in 1914/5 rising to 300–350,000 tons a year from 1917/8 onwards, with the option of negotiating for even greater quantities.

The second point, about government control, came up once negotiations were reopened early in 1914. The government proposed to appoint two directors, armed with a veto over any board decision. Cargill protested that such a veto would give future governments the power to interfere in the day-to-day running of Anglo–Persian for the sake of short-term political advantage.

In consequence a Treasury letter, written also on 20 May, restricted the veto to questions of general policy such as high matters of state, or any attempt by the company to sell out, particularly to foreigners, or to change its status or the disposal of oil or products in such a way as to jeopardize the fulfilment of Admiralty contracts. The two *ex officio* directors appointed were Admiral Slade and a ship-owning magnate with a good knowledge of India and the Persian Gulf, Lord Inchcape.

By a fluke, the bulk of the money to meet the government share acquisition was already available in the Exchequer's sinking funds, for meeting "belated naval payments". The diversion of these funds to a different purpose needed parliamentary sanction. The main House of Commons debate took place on 17 June, Greenway sitting in the official box alongside

senior Treasury men in case Churchill should require specialized information during the debate. Cargill had asked for a puff
to be included in Churchill's speech, praising Burmah Oil's
kerosene price policy in India; his wish was not met.

That debate, noteworthy for the exchanges between Churchill and Sir Marcus Samuel's MP brother, was not the only
occasion on which Shell's spokesman actively challenged the
Anglo–Persian agreements. At the annual meeting of one of his
companies the previous month, Sir Marcus had questioned the
Admiralty decision to subsidize a company operating in an
insecure foreign country at the very moment when supplies in
British territory under British jurisdiction were "making rapid
and most satisfactory progress". (However, none of the countries he cited – Egypt, Trinidad and Sarawak – produced nearly
such good results as did Persia.)

Again, the *Petroleum Review*, edited by one of Shell's consul-
tants, published a series of articles attacking both the government and Burmah Oil over the agreement, and Greenway
accused Cohen of having supplied the information on which
those articles were based. Cohen convincingly disproved this,
adding that so far from promoting the attacks in the press, he
and his executive colleagues (which demonstrably included
Deterding but not Samuel) were concerned not to let the
Anglo–Persian agreement sour their relations with Whitehall.

A short while later Cohen wrote to Finlay, offering a small
gesture of good will to Anglo–Persian, adding, "It will be a
great pleasure to me to celebrate in that way the sealing up of all
the 'rifts'." Even Cargill managed to extract from the *rapprochement* some modicum of cheer. He noted with superlative caution, that "Anglo–Persian Oil Company's outstanding questions have been pretty well settled with Cohen, and on the
whole it seems to me that, everything considered, arrangements
made are of a fairly satisfactory nature". They were not to
remain satisfactory for long.

Notes

1 An interesting and on the whole accurate account of the events in these
   years, based mainly on the public records, is in M. Jack, "The Purchase of
   the British Government's Shares in the British Petroleum Company (*sic*)
   1912–1914", *Past and Present*, 39, 1968, pp. 139–68.

2 Admiralty Committee on Use of Fuel Oil in Navy (see ch. VII, note 5). Deterding and Samuel were interviewed on 21 December 1911, and Strathcona and Greenway on 29 December 1911.

3 PRO ADM 116/1208: Cargill being interviewed on 29 October and Greenway and Samuel (separately) on 19 November 1912.

4 PRO FO 371/1486: memo L. Mallet to Sir A. Nicolson, 29 or 30 August 1912.

5 ibid., Admiralty to Foreign Office, 26 September 1912. This important file, which deals also with Mesopotamian affairs, continues to 28 December 1912, FO 371/1760–1 covering 1913 and FO 371/2120–1 covering 1914.

6 See letters Churchill to First Sea Lord, 8 January and 8 May 1913, R. S. Churchill, *Winston S. Churchill*, Companion volume II, Part 3 1911–1914, (1969), pp. 1932–4 and 1941–3.

7 PRO CAB 37/115, Appendix 2: Admiralty Memorandum in Regard to Anglo–Persian's Proposals.

8 ibid., 16 June 1913: Churchill's supporting memorandum "Proposed Arrangement with Anglo–Persian Oil Company for Supply of Fuel Oil", 4 July (CAB 37/116), is also in R. S. Churchill, *Winston S. Churchill*, Companion Volume II, Part 3, pp. 1945–7.

9 Marder, A. J. (ed.), *Fear God and Dread Nought III 1914–20* (1959), p. 338. Marder himself, in *From the Dreadnought to Scapa Flow: I The Road to War 1904–1914* (1961), p. 407, commends Slade as a "keen student of war", cf. ibid., V (1970) p. 321.

10 Command Paper Cd 7419 (1914): "Agreement with the Anglo–Persian Oil Company, Limited", Final Report of Admiralty Commission on the Persian Oilfields, para. 30, p. 27.

11 The legend has grown up that Greenway was incompetent as chairman of Anglo–Persian (see Sampson, *The Seven Sisters*, p. 55). Since appearances count for so much in this publicity-conscious century, the damaging sobriquet of Upton Sinclair, in his not very good novel *Oil!* (1927), "Old Spats and Monocle", has helped to set the image of Greenway (see Plate 28). However, as the Burmah Oil archives and public records show, he was a very persistent and single-minded bargainer, who above all won over the British government to accepting very good terms (on both sides) for the acquisition of a majority shareholding in Anglo–Persian. As Dr Ferrier's article, "The Early Management Organization of British Petroleum and Sir John Cadman", in L. Hannah (ed.), *Management Strategy and Business Development* (1976), especially p. 136, makes clear, Greenway was not the organization man that his successor Cadman was, but did have the energy to see the company through its difficult early years, set up a tanker company in 1915 and later a marketing company by acquiring British Petroleum, and later still moved it to a world-wide role.

12 Command Paper Cd 7419 (1914), pp. 15–22, with explanatory memorandum, pp. 3–7. The contract was not published at the time and the Treasury letter only in 1929. See Hansard 5th Series, Vol. 226, 1928–9, cols 2263–4 (26 March 1929).

CHAPTER XIII

The Mesopotamian Concessions[1]

In Iraq, a mile or so away from Kirkuk's "eternal fires" popularly associated with Nebuchadnezzar, a jet of oil shot so high into the sky in October 1927 that it could be seen twelve miles away. Baba Gurgur was the site of the Middle East's second great oil discovery. It was made by the Turkish Petroleum Company (later the Iraq Petroleum Company) and more than 12,000 tons of oil drenched the surrounding countryside daily until the flow was contained.

Burmah Oil had been directly concerned in this venture from its beginnings in 1905. Even when the early negotiations were taken over by Anglo–Persian in 1909, Cargill and his colleagues retained a measure of responsibility for financial commitments while Anglo–Persian was making no profits, and subsequently as directors of that company. By 1927, Anglo–Persian's stake in the Iraqi concessions was just under 25 per cent.

Before the First World War Mesopotamia (as it was then called) was an outlying part of the Turkish empire. In the northern and central provinces of Mosul and Baghdad there were certain oil indications, notably at Kirkuk. W. K. D'Arcy started bidding for a concession there in 1901 as soon as he had secured the Persian one. There were two reasons why he was anxious to add another 120,000 square miles or so to the 500,000 over which he then had rights. If oil were discovered in Persia at Chiah Sourkh, its transportation would be less costly down the river Tigris to the Persian Gulf, and later by a future pipeline along the plains of Mesopotamia, than over the Persian mountains. Moreover, as Chiah Sourkh was not far from the as yet undefined frontier between Turkey and Persia, the two concessions worked together would yield substantial operational economies, whereas there could otherwise only be constant friction between rival companies in such close proximity one to the other.

Redwood therefore arranged for his geologist H. T. Burls to visit Mesopotamia and carry out a survey similar to the one he had undertaken in Persia. That survey encouraged D'Arcy to send H. E. Nichols, a manager in his gold mining company, out to Constantinople in 1903 to negotiate. However, in contrast with the relatively easy way in which the Persian concession had been won, Nichols was to find the going extremely tortuous amid the Byzantine bureaucracy and intrigue at the sultan's court.

Sure enough, finding himself competing against such German interests as the Deutsche Bank, which controlled the Anatolian and Baghdad railway companies, Nichols made little progress in the period before Burmah Oil's directors and D'Arcy were brought together in August 1904. Under an agreement of 20 May 1905 the company stood ready to purchase all D'Arcy's rights in Turkey as well as Persia. The Concessions Syndicate then took charge of the negotiations in Constantinople and arranged to share with D'Arcy all subsequent out-of-pocket expenses. They left open the question of financing the production, transportation and refining of any oil that might be discovered.

Predictably, Cargill much resented Burmah Oil becoming involved in another such project, but reluctantly went along with the sanguine views of Redwood and others. Not until the end of 1906, when the German bankers through inaction had forfeited any rights to the concession, did the Glasgow board for the first time seriously consider Mesopotamia. It authorized Cargill, Wallace and Hamilton to carry negotiations to the stage of a firm offer which the full board would then consider. Should a decision have to be taken in a hurry, the three could act on their own, but must not commit Burmah Oil to more than £50,000 in cash, nor to substantial drilling or development expenditure.

Mesopotamian affairs were thus being transferred to an *ad hoc* sub-committee, effectively controlled from London by the strong-minded Wallace. Then Nichols in Constantinople suddenly found himself caught up in a flurry of activity that seemed to bring the concession almost within his grasp. To the Burmah Oil people, that was by no means the glad news it should have been; not only had D'Arcy refused Cargill's demand that he should bear half of future expenses at a time when funds ear-marked for Persia were running low, but the three directors were also far from unanimous on what to do. Hamilton, the recipient of increasingly urgent cables from Nichols, was all for instructing him to go ahead without delay, so as to prevent the concession from being snatched away by someone else. Wallace and Adamson – deputizing for Cargill, absent from Britain for health reasons – sought on the other hand to drag out the negotiations for as long as possible, convinced that delay was the best course while so little was known of the likely costs.

Wallace put Hamilton and his impatience in their place with a splendid piece of condescension, writing as follows to Adamson:

> Hamilton has never been to the east and does not understand the beautiful policy of dawdling. Tell him you do not need a "loophole" for dawdling; you just dawdle – leave letters unanswered for a day or two, when pushed answer that you will reply as soon as possible, that you are consulting so and so, ask various questions etc. etc. etc.

> There are many reasons for this, and as Fleming agrees and, I believe, you also agree with me, Hamilton must consider himself outvoted, as I have so often been. And let him take a leaf out of my book. When outvoted, a good man makes the best of the policy advocated by the majority. . . . In short, we come back again to "just dawdle" and writing to an old Indian [hand] like you, I don't have to explain what I mean.

Hamilton, still very much a new boy on the board, happened to open that letter. "True I have never been east," he replied rather huffily, "but do understand a little about dawdling": clearly a dig at his colleagues' reluctance to give a straight answer to importunate messages from Nichols such as, "We cannot play with our Turkish friends much longer." Indeed, Hamilton went on with defiance, "so far as being outvoted is concerned, it will not be the first time, and having done what I consider my duty, I have no feeling in the matter, although I consider that I am right and the others wrong."

Despite this dithering from home, Nichols did succeed in inducing the Turkish authorities to open preliminary talks early in 1907. But he was all too aware that rival bidders might well offer more attractive terms, including promises of copious ready money. A number of such tempting offers were indeed made, but Hamilton, by then converted to a wait-and-see policy, reassured D'Arcy that the rivals were "novices as far as oil is concerned – I think we should let them have a try," he added. "If they attempt to work the fields themselves, I am of the opinion that they will require to come to us before they are very much older": a familiar story among concession-owners in Burma. One by one those rivals began to lose interest when they came up against all the maddening delays, and left Nichols as the only serious contender.

Reynolds's strike of oil in Persia in May 1908, while it magically restored good relations between D'Arcy and his Burmah Oil colleagues, did nothing to resolve the problem of Mesopotamia. D'Arcy himself was in two minds, apprehensive about losing the concession to rivals, once again on the alert now that the Persian success was being "noised abroad", but feeling quite unable to accept the company's requirement that he should meet half the eventual costs. Perhaps, he suggested, some outside "capitalist" might be induced to set up a financially and operationally independent company for Mesopotamia in which the prospective Persian company, and hence Burmah Oil, would have a minority interest.

That scheme evoked no enthusiasm when put to the Burmah Oil directors, who above all gloried in running their own show. Wallace and Fleming argued powerfully that D'Arcy, egged on by his "capitalist" friend, might in such a case subsequently turn against his former colleagues and "twist the tail" of Burmah Oil as part-owner of any Turkish company, which could only make trouble for the Persian company. If, instead, both sets of concessions were firmly in Burmah Oil's hands, then there could be no border disputes and the concessions could be worked jointly in harmony. The Glasgow board accepted this argument and agreed to allow the Concessions Syndicate directors to bid for the Mesopotamian rights and spend up to £150,000 of Burmah Oil money. On 19 August a Burmah Oil agreement with D'Arcy recognized the parlous financial circumstances which he constantly pleaded, by not requiring him to contribute for the time being but allowing him to acquire a 40 per cent interest in any concession obtained – presumably out of his eventual proceeds from Persia.

The preparatory discussions between Wallace, D'Arcy and their respective lawyers included a moment of farce that was probably not all that rare in the life of a preoccupied, pleasure-loving and in some ways slightly woolly man such as D'Arcy. Whenever he had found himself short of cash, he had reckoned to form *ad hoc* syndicates to finance different types of ventures. Now his solicitor unexpectedly produced an old agreement and asked whether he remembered it. "It made D'Arcy hot all over," Wallace gleefully reported to Cargill. "He had forgotten part of its effect, which was that [certain associates in the original Persian venture] were to get between them one-tenth of the Turkish concession. Tableau. Poor old D'A. He was bled by everybody!!" After that little difficulty, he was made to give the company an undertaking that he would pay any such outside beneficiaries out of his own future returns. Wallace in passing rubbed it well in that he "had been much too generous to a great many people".

Nichols, now an employee of the Concessions Syndicate, was in February 1909 offered a contract by the Turkish government for the Mesopotamian concession, the terms being comfortably within the limits set by the Burmah Oil board. Even so, Nichols hoped to do even better and wrote asking Hamilton to travel out at once to act as plenipotentiary. Wallace without delay instructed Greenway and James Thompson, the company's solicitor, to go as well and in the space of a morning hammered out the precise terms with D'Arcy and his sundry legal advisers. A formal agreement, concluded on 25 March between Burmah Oil and D'Arcy, cancelled that of August 1908; now D'Arcy sold his entire rights in Mesopotamia in exchange for 10,000 £1 shares of Burmah Oil. That complemented an accord of the same day over Persia (see Chapter IX). He also undertook to lend his name to any initiative that might help obtain the concession.

Any enthusiasm Hamilton and his colleagues may have had for catching the Orient Express soon waned, as they were all frantically busy setting up the Persian company; in the end none of them could find time to make the trip. Nichols on his own strove to secure the contract which, by an unusual burst of official energy, was agreed and ready for signature by 13 April, 1909. Then on the following day, when the palace was due to authorize its signing, the Young Turk revolution broke out in Constantinople and the sultan was driven into exile. Nichols returned home as the capital was in too much turmoil for any business of substance to be transacted.

On 14 April, too, Anglo–Persian had been established. Its prospectus did not mention Mesopotamia by name, although the two agreements of 25 March were listed among the contracts binding the new company. But rivals were springing up to press their own claims to the concession, and those included the Shell group. In July Sir Marcus Samuel wrote to the Foreign Office and requested its support for Shell in Constantinople, but the Foreign Office made it clear that D'Arcy already had a prior claim.[2] The fact that D'Arcy now had Foreign Office backing gratified the Admiralty, which accepted that his financial sponsor, Burmah Oil, deserved some official consideration – but preferably not at the taxpayer's expense – for having supplied naval fuel oil from Rangoon at great trouble and cost.

Later in 1909 Wallace proposed that responsibility for Mesopotamian negotiations should be transferred from Burmah Oil to Anglo–Persian, where logically it now belonged. Greenway, about to take over as managing director, wondered whether Anglo–Persian's finances would run to it, as £100,000 would probably have to be spent there before any returns could be expected. For the time being, therefore, the actual financial burden would continue to fall on Burmah Oil, although in October 1909 Anglo–Persian took over Mesopotamian affairs and repaid to the company the total outlay to date.

Despite this potential commitment looming over himself and the Glasgow board, Cargill was keeping his pecker up by hoping that they would eventually lose the concession! Then towards the end of 1911 a formidable opposition group emerged, sponsored by Gulbenkian, and comprising the Deutsche Bank, Shell and the British-owned National Bank of Turkey: it was later incorporated in London as the Turkish Petroleum Company. For Anglo–Persian, by then greatly embarrassed by a shortage of funds, that was bad news. Cargill vetoed any move by Burmah Oil to "spend money on Turkish ventures", so that Anglo–Persian's board merely agreed to consider any proposals by Greenway not involving the use of finance – a virtual ban on future initiatives. Greenway had other ideas: in July the Foreign Office informed him of reports from the embassy in Constantinople that the Deutsche Bank–Shell consortium was actively in the fray. He at once drafted a reply, committing Anglo–Persian to taking up the concession wholly on its own.

On receiving a copy of the draft, Cargill's anxiety and irritation boiled over in the letter already mentioned in Chapter XII:

> You say that there would be no difficulty in forming a syndicate to take up the [Mesopotamia] concession should this prove necessary, or in finding any portion of the capital which the Burmah Oil Company/Anglo–Persian Oil Company are unable or do not desire to take. . . . How would you propose that the Anglo–Persian Oil Company should find the capital for this purpose? Before long they will be considerably in debt, and as far as I am concerned the Burmah Oil Company will certainly not guarantee interest on any more Anglo–Persian Oil Company preference shares to develop a concession which under the most favourable circumstances would not be revenue, let alone profit, earning for a very considerable period. Nor as regards the Burmah Oil Company would I feel prepared to recommend that company putting any capital into such an enterprise.
>
> Your real object is of course to prevent [outsiders'] competition with the Anglo–Persian Oil Company, but after all that competition would only be likely to seriously affect the local, i.e. Turkish and Persian, trade of the Anglo–Persian Oil Company, and it remains yet to be seen what that trade is likely to amount to. Competition from Turkey is not likely to affect in any way the arrangement we have made with the Asiatic Petroleum Company in regard to kerosene and benzine, nor would competition from Turkey have the slightest effect on our fuel oil trade, which is the direction on which I understand you are pinning your greatest hopes for the future of the Anglo–Persian Oil Company.

Cargill was anxious to drive home three points. First, Burmah Oil was not prepared to throw away any more money in Mesopotamia. Second, the Mesopotamian concessions, even if eventually productive, would probably be no more than a limited safeguard for Anglo–Persian's activities. Third, he himself had constantly to worry about finance and the reactions and interests of his many shareholders, whereas Greenway had no ordinary shareholders as such, and meanwhile could pay his preference dividends only because these were guaranteed by Burmah Oil.

Greenway was therefore left in no doubt as to what his company's ultimate paymaster thought of Mesopotamia. Yet that did not stop him from taking a tough stance at a meeting held shortly afterwards at the Foreign Office.[3] There he spoke his mind very freely. It had been on imperial grounds that the Admiralty had pressed Burmah Oil to take up the Persian concession, he explained, and on the same grounds Anglo–Persian had lately rebuffed every overture by the Shell people. Yet if Shell and its consortium partners were to gain Mesopotamia, Anglo–Persian would inevitably be forced to come to terms since, he argued, the eventual flow of oil from there would be "overwhelming" and (with some hyperbole) practically the whole of the world's oil supply would be controlled by foreigners, much of it by the Shell group.

As it seemed the practice of Greenway and his friends to join their foes in agreements if they could not beat them, the Foreign Office suggested that Anglo–Persian might voluntarily team up with the consortium and take a quarter share. Greenway dismissed that as unthinkable unless Anglo–Persian were allowed to manage and control the Turkish Petroleum Company's business and a British chairman, with a casting vote, were appointed. He conceded that some limiting of competition in the middle east would follow so that, if the Admiralty objected, there would be no question of going into the consortium.

The Admiralty was therefore approached by the Foreign Office and took the grand strategic line: Persia's oilfields must, come what may, remain in British hands, and Whitehall must give no encouragement whatever to foreign interests anywhere in the region. When Admiralty officials opened with Greenway discussions on fuel oil supplies from Anglo–Persian that November, they talked only of Persia, and it was the Foreign Office which turned down a request by the Turkish Petroleum Company for diplomatic support over getting the Mesopota-

mian concessions: a move which predictably delighted Greenway when reported to him.

To try to drive home to the authorities what he alleged was
the untrustworthiness of the Shell people as business partners,
Greenway had a dossier of correspondence with Cohen printed,
and sent a copy to the Foreign Office.[4] The letters were of the
usual knockabout type that would have raised no eyebrows in
the Burmah Oil or Anglo–Persian sanctums, and in this case
concentrated on the very sensitive area of fuel oil supplies. They
showed Cargill, Greenway and Hamilton meeting Cohen and
his associates and coming to a verbal agreement over surplus
fuel oil not required by the Admiralty, and then Cohen a day
later claiming that their accord had been much more limited in
scope than the written confirmation had indicated.

Greenway and Cohen next bickered for a couple of days
about exactly what had been agreed, after which Cargill entered the contest on Greenway's side. Cohen, who had probably been put up to it by Deterding, was unable to break the
united front against him and gave in. On the evidence of the
dossier, Greenway pressed his case that Shell was striving to
achieve complete control over the Admiralty's fuel oil supplies.
His astonishing success appears from the minute by the Foreign
Secretary, Sir Edward Grey, when the file was referred up to
him:[5]

> We are to continue to make every endeavour to obtain the Mosul
> and Baghdad fields [in Mesopotamia] for the Anglo–Persian Oil
> Company and if we eventually fail, His Majesty's government must
> consider the question of subsidizing the Anglo–Persian Oil Com-
> pany to prevent its absorption into a trust. Of course, this last
> alternative must not be mentioned to Mr. Greenway.

Cargill had been shaken by that row with Cohen about fuel
oil, and feared lest the surplus stocks might have to be destroyed
at the refinery if there were an open quarrel with the Shell
group. Then in January 1913, on the same day as he wrote his
doleful letter to Finlay quoted in Chapter XII, he bared his
heart almost as candidly to Adamson:

As I said from the very start we had neither the *organization* nor the *number* of experienced and practical men necessary to properly develop and run the business [in Persia], and it is that that is largely to blame for the awful mess things are now in. And yet here we are contemplating taking on this *huge* Mesopotamian concession and developing and working it!! Again I say, *where are the men* to undertake it with any hope of real success?! I *can't* help worrying myself over all this and wondering what is in store for me when I get home.

After all, he asked, would it be such a terrible thing if the Shell group were to secure the Mesopotamian oil and use it against the company, now that Asiatic had agreements to market most of Anglo–Persian's products? To him, the best outcome would be for nothing to happen. "The more I think of now attempting to tackle this [Mesopotamian] concern – except absolutely on the terms of hanging it up and doing no active work on it – the less I like it!"

Fortunately for Cargill's fragile peace of mind, little progress was being made in Constantinople. When officials there hinted that a payment of £50,000 to the Turkish government might be enough to gain the concession, Burmah Oil agreed to provide that sum, half directly and half as a loan to Anglo–Persian. Nichols, now regarded by officials as having been out there too long and hence "quite useless", was brought home and replaced by a locally based man. Yet the atmosphere of dawdling in the capital persisted, with the Turkish authorities reluctant to break the deadlock by coming down on one side or the other.

Meanwhile in Whitehall, Sir Edward Grey held a top-level meeting, attended by the permanent secretary of the Board of Trade as well as Greenway. True to his earlier minute, Grey was seeking to promote a D'Arcy–Turkish Petroleum consortium, but on Greenway's terms of at least 52 per cent financial stake, with a British chairman having a casting vote and marketing arrangements that would give the D'Arcy group rights over half the refined products. However, the Foreign Office found the gap between the D'Arcy group and Turkish Petroleum too wide. In July 1913, probably in deference to views attributed to the Grand Vizir of Turkey, it made a startling change of direction and proposed opening talks with the German government for the British interests and the

26, 27. Visit of the Viceroy and Lady Chelmsford to the Yenangyaung field, 1916.
Top: Irrawaddy Flotilla Company express paddle steamer *Taping* with the
Viceregal party. Bottom: The motorcade at the oilfield.

28. Charles (later Lord) Greenway (1857–1934), chairman of the Anglo-Persian Oil Company 1914–1 (Courtesy of Mrs. Dalzell Hunter)

29. The air age reaches India. Karrier aircraft-fuelling lorry, Calcutta c.1921.

30. Sir Henri Deterding (1866–1939), managing director of the Asiatic Petroleum Company. (A Shell photograph)

31. Elephants bringing in the boiler to provide power at the wells, Burma 1915.

32. Absent friends. Finlay Fleming & Company assistants celebrating Christmas, 1921.

33. Sir Robert Waley Cohen (1877–1952), director, the Asiatic Petroleum Company. (A Shell photograph)

Deutsche Bank to share the concession, keeping out the Shell interests altogether.

This *volte face* clearly recognized that the D'Arcy group had no chance now of securing a dominant role in Mesopotamia. Instead, here was an opportunity of settling an outstanding point of friction between Britain and Germany in that particular area of rivalry, the Turkish empire. At the same time, as Greenway was told by the Board of Trade, both Whitehall and Berlin had common aims in isolating and if possible offsetting Shell. The German government was anxious to break the monopoly of Shell and Standard Oil in Europe by developing a market for refined products from German-financed output, and had no objections to the D'Arcy group having technical control over the operations. Any such attempt to exclude Shell was, however, finally knocked on the head when Deterding, whom Churchill had summoned to the Admiralty to see if he could speed up deliveries of naval fuel oil under the 1913 contract, complained to the First Lord that he had been shockingly betrayed by the Foreign Office. Grey had no chance of winning this battle with the persuasive Churchill, and Shell was restored to a place in the consortium.

On 19 March 1914 the British and German governments and the other parties concluded an agreement at the Foreign Office, Greenway and Barnes signing for the D'Arcy group. The Turkish Petroleum Company was to have its capital doubled to a figure of £160,000, and would become responsible for exploring and proving oilfields in Mesopotamia. The D'Arcy group would have a 50 per cent share in Turkish Petroleum, as well as four seats on the eight-man board, while Shell and the Deutsche Bank shared the remainder, although they both gave 2½ per cent of their shares to Calouste Gulbenkian. Like the kerosene agreement of 1905 between Burmah Oil and Asiatic – two parties contending in the same market and deeply suspicious of each other – the 1914 agreement reflected a drawn battle between rivals rather than a cordial alliance of like-minded people. Although many items remained for further decision, such as the exact form of the producing and marketing organization and the shares of products that would fall to each party, Turkish Petroleum could at last be put on a sound footing.

The chairman of that company was to be nominated by the

D'Arcy group, but subject to Whitehall approval: he should be a British national and a man of eminence. In April the Anglo–Persian board, at which Wallace but not Cargill was present, proposed Sir Hugh Barnes as chairman and Greenway and D'Arcy as two of the other nominees. On hearing the news, Alwyn Parker of the Foreign Office[6] exploded in a confidential minute that Barnes was far from eminent and to Parker's knowledge "remarkably dense and slow at the uptake". Barnes was neatly jettisoned by being told that it was essential for the future chairman to have no links whatever with Anglo–Persian.

Greenway then put forward the name of a Scottish laird with two colonial governorships behind him. That went down even more badly with Parker, who minuted, "I could hardly imagine a worse choice. He would be quite unable to stand up to people like Deterding and Stauss" (of the Deutsche Bank). His name was dropped, and the search continued without result until the outbreak of the First World War that August made the question academic.

A no less important question was how the D'Arcy group would finance its £80,000 half-share in Turkish Petroleum. Greenway told his board that the money could be put up by an all-British holding company, with a capital of £100,000: the additional £20,000 would cover the expenses of preliminary tests. The holding company was to be called the Stanmore Investment Company "in order not to excite curiosity", as he explained to Black at the Admiralty. Stanmore Hall in Middlesex was one of D'Arcy's residences, but the name had the quite misleading aura of a development company to construct villas in a pleasant outer-London suburb; shortly afterwards it was more felicitously rechristened the D'Arcy Exploration Company.

Whitehall was almost over-fussy in its instructions over the source of the holding company's finance. Whereas the Admiralty presumed that it would be subscribed directly by Anglo–Persian, the Foreign Office insisted that it must come from "the D'Arcy group as distinct from the Anglo–Persian Oil Company." That attempt to make a distinction without a difference occurred because the British government was in the summer of 1914 on the point of acquiring its majority shareholding in Anglo–Persian, so that foreign powers might be suspicious of the consequent indirect link between Whitehall and the Turkish concession that had still not been secured. The Board of Trade sought to cut the knot with the more realistic suggestion that the independent Burmah Oil should be the source of funds.

That brought the cautious Cargill back into the debate. He refused to accept the proposal unless the actual beneficiary, Anglo–Persian, agreed to indemnify Burmah Oil against any financial loss should no oil be discovered in Mesopotamia. The Treasury did not care much for that condition, and tried to restrict Anglo–Persian's indemnity to no more than half of any loss. Greenway explained to officials that Burmah Oil – he really meant Cargill – had all along had little desire to extend its interests into Mesopotamia, and would in any case bear a third of any potential loss through its shareholding in Anglo–Persian. The Treasury was so dissatisfied with this response that it pressed for the setting up of a completely independent consortium, unconnected with Anglo–Persian. The Foreign Office added its voice by stressing the German and Turkish governments' deeply held suspicions of Anglo–Persian involvement in Turkish Petroleum, and made the alternative proposal that D'Arcy should become chairman and principal shareholder of his exploration company.

That broad identity of interest between the Treasury and Foreign Office, in wishing to erect a *cordon sanitaire* between Anglo–Persian and the Mesopotamian venture, chimed in with Cargill's insistence – by then almost an obsession – that D'Arcy must personally provide a quarter of the exploration company's capital, even though Burmah Oil would have to lend him the money on the security of his shares in the company. Then, before Turkish Petroleum could secure the concession in Constantinople, war broke out with Germany in August 1914, to be followed by the entry of Turkey on the enemy side that November.

Britain and her major allies in the entente were right in thinking that the war would finally bring down the ramshackle Turkish empire, and all had a direct interest in its demise. Russia, its way into Persia blocked by D'Arcy, was looking to Constantinople and the Straits, France to strengthening her traditional hold on the Levant, and Britain to setting up friendly protectorates further east to safeguard her communications with India. The revolution of 1917 for the time being removed the

Russian threat to the area, and a new element was the prot-racted negotiations – of marginal interest to Burmah Oil – between Britain and France over Mesopotamia.

The Harcourt-Deterding accord of April 1919 (see Chapter XVI) neatly tied up two loose ends that concerned the company: an arrangement whereby Anglo–Persian and Shell should participate on equal terms in Mesopotamia, and the parallel one whereby the Shell group, apart from its wholly Dutch interests, should come under British control. Although, as Chapter XVI makes clear, Shell never became entirely British, the Mesopotamian clauses basically held, but both parties saw their shares scaled down. A year later the British and French governments finally agreed that France should have the 25 per cent share previously given to the Germans.

By then United States interests, alarmed by dwindling oil supplies at home and by Britain's apparent hogging of potential supplies elsewhere in the world, demanded a stake in Mesopo-tamia, now Iraq. That demand could be justified by the fact that American oil companies had, during the war, over-depleted their reserves to keep the allies in Europe supplied. Those companies, backed by the administration in Washing-ton, were now tireless in advancing their case. It took until 1928, shortly after the Baba Gurgur discovery mentioned at the beginning of this chapter, before a definitive settlement gave Anglo–Persian, Shell, the French and the Americans just under 25 per cent each, and Gulbenkian his celebrated 5 per cent.

Meanwhile, about the only immediate concern of the Bur-mah Oil directors was the rather low-grade one as to why D'Arcy, and after his death in 1917, his executors, should have been allowed a share – amounting to no less than 12½ per cent – in the Mesopotamian concession when he had so patently signed away all his interests in 1909. That was apparently an act of generosity by Greenway, who accepted the executors' right to such a share. For years afterwards the very pertinacious R. I. Watson was to mull over the deplorable sequence of irrational acts and misjudgments that had quite wrongly, in his view, restored to D'Arcy a stake in Mesopotamia.

Notes

1 Two essential outside sources are L. Lockhart's *Anglo–Persian Oil Co.*, XVII and XXX, "The Turkish (Mesopotamian) Concession Negotiations', 1901–1909 and 1909–1914 respectively, and M. Kent, *Oil and Empire: British Policy and Mesopotamian Oil 1900–1920* (1976).

2 See series of letters and memoranda in PRO FO 371/777, starting with Samuel to Sir E. Grey (Foreign Secretary), 28 July 1909, and ending with memo Admiralty to Foreign Office, 28 October 1909.

3 See reference in ch. XII, note 4.

4 PRO FO 371/1486: Greenway to Maxwell (Foreign Office), 30 October 1912, enclosing printed copy of correspondence between Anglo–Persian and Asiatic, 15–24 October 1912.

5 ibid., minute by Sir E. Grey, 11 November 1912, following memo by L. Mallet, 6 November.

6 Alwyn Parker was a middle-ranking official in the Foreign Office, who seems to have been the expert on oil questions. He resigned in 1919 when only 42; presumably his outspokenness about Anglo–Persian and its top management prevented him from joining the procession of ex-public servants into that company as ordinary directors or senior managers. These included Sir Hugh Barnes, Sir Frederick Black, Sir John Cadman, J. C. Clarke, Sir Edmond Slade, Sir Francis Hopwood (Lord Southborough) and Sir Arnold Wilson. Sir George Barstow, Sir John Bradbury, Sir Walter Jenkins and Sir Edward Packe were exclusively government-appointed directors. Parker's memorandum of 8 May 1914 about the chairmanship of Turkish Petroleum is in FO 371/2121; he was one of those who sought to draw a distinction between the D'Arcy Group and Anglo–Persian. His comment about the Scottish laird (Lord Lamington) is in ibid., 14 July 1914.

CHAPTER XIV

Reform in the London Office
1912–14

The euphoria of the Burmah Oil directors, caused by the Bank of Burma's collapse, all too soon gave way to a mood of more sombre realism and the conviction that now was the time to patch up relations with British Burmah. Preliminary feelers led to full-scale talks. At first, each side was deeply mistrustful of the other, but Cargill soon came to respect the integrity of the British Burmah people. Now that Mower and Clifford were behind bars and about to be sentenced to terms of imprisonment, he told Finlay that their very different successors could be relied on to collaborate in a straightforward and friendly fashion.

An accord was signed in May 1912, agreeably settling all points at issue between them. It renewed the arrangements with Rangoon Oil for crude oil deliveries, but on a flexible basis as British Burmah needed a certain quantity to keep its Thilawa refinery going. Burmah Oil was to take over the marketing of all its refined products. For once in his life Cargill expressed unalloyed satisfaction, writing jubilantly to Rangoon, "I fancy the Indo–Burma Petroleum Company will be pretty sick when they hear that we have fixed up the British Burmah Petroleum Company." Indeed, one of his motives for seeking a *rapprochement* had been to counter the energetic efforts by Duncan of Steel Bros to arrange a merger between Indo-Burma and British Burmah.

Cargill anticipated other important bonuses for the company. For instance, he felt that the governments of India and Burma would look more favourably on it once it was seen to be

214

working harmoniously with its most important rival. Moreover, now that it no longer needed to offset British Burmah's wells, its hands would be strengthened in dealing with the extravagant working costs in the fields. Those costs were once again giving him great anxiety. "The amount we are spending annually on field work is simply appalling," he wrote sternly to Rangoon, "and every effort must be put forward to reduce expenditure in every direction." He was especially upset because the fields agent was calling for large increases in the number of drillers, something he unhesitatingly turned down.

Since the new agreement would have an impact on those already existing with Asiatic, how would Cohen react? He had been away in Egypt securing concessions clearly intended to offset those of Persia, which led Greenway to use a chess metaphor of the "Asiatic Petroleum Company's Egyptian knight brought out to defend the attack of our Persian queen." Immediately after Cohen's return Cargill saw him; at first Cohen conceded the value of such an agreement, but very soon – presumably having reported back to Deterding – complained that British Burmah would use it as a pretext for pushing up its share of the Indian kerosene trade. Under the terms of existing arrangements, he claimed that the burden would fall on Asiatic. Cargill and Greenway at once offered to meet him half-way; as usual a "tremendous amount of haggling" ensued. Cohen soon simmered down and in a supplementary memorandum confirmed that Asiatic was a full party to the agreement's terms.

Indo–Burma was now left as the only unfriendly competitor of any consequence in the province, and still an irritant, since it was making unwelcome inroads into the kerosene trade of Chittagong and Assam. However, Cargill argued and Cohen agreed that, all in all, the only rational policy was to keep Indo–Burma at arm's length.

Cargill was still smarting over the ease with which Standard Oil had augmented its market share in India during the recent price war, and in the autumn of 1911 he had asked Wallace and Greenway to draft a memorandum for the India Office and the government of India, detailing the company's difficulties. That memorandum claimed that Asiatic and Standard Oil, although ostensible adversaries in the price war then under way, in practice had identical objectives: to destroy indigenous companies in India and Burma "by a process of extinction and absorption". The local governments should therefore step in and preserve both industries and consumers from "the evil consequences which would follow upon an American /Dutch monopoly."

Before the memorandum could be submitted, however, persistent rumours reached the company that the government of India would be imposing an excise duty on domestic kerosene production, as had been feared in 1910 (see Chapter XI). The directors at home therefore suggested that Finlay – already a member of the Burma legislative council in Rangoon – should have himself nominated to the viceroy's council, which happened to have a vacancy, with the object of "fighting the excise question". Finlay was not at all keen, but consented as long as that question was definitely on the agenda.

Then Cargill changed his mind and contented himself with submitting to the authorities a note showing that the existing oil royalty, in absolute terms and as a percentage of the company's earnings, was heavier than anywhere else in the world. He and Greenway discussed the note with Sir Thomas Holderness at the India Office. Holderness pointed out that Indian civil servants still regarded Burmah Oil as making excessive profits; in answer to Cargill's insistence that oil was a wasting asset and production already in decline, he rather heartlessly declared that "if Burma oilfields were going to pan out in a comparatively short time, the government need not worry their head about protecting the industry in any way."

After Cargill once again directed Finlay to go, and then cancelled that direction, Finlay's anger exploded. "I find it difficult to carry on if you change your instructions in this way," he declared, and brought up some other grievances where he felt he was receiving scant support from home, notably over Lim Chin Tsong (see Chapter X). It illustrates the close relationship between Burmah Oil and its agency that Cargill at once unreservedly backed down, expressing his sympathy over Finlay's troubles in Rangoon. He had recently had a further and much more satisfactory session with Holderness, who had clearly been briefing himself about the company's problems and seemed to be far more understanding. Finlay did not in the end join the viceroy's council, and no excise duty was levied for a decade or so.

Cargill later explained to Finlay why he had dithered so much over the proposal. In particular, the split operations between London and Glasgow caused him any number of difficulties:

> Were the managers and directors all concentrated in the one place it would be a comparatively simple matter to arrive fairly promptly at final decisions on all matters of importance, but when one has to attempt to do this by correspondence and when the London directors are scattered and some of them too busy with other things to give immediate attention to questions I wish to consult them about, it is no easy matter sometimes to settle these questions promptly and satisfactorily.

Here then was the essence of the company's top management troubles. The London end was run by Greenway who was also the managing director of Anglo–Persian, already a large company in its own right, and who never got on well with Cargill. Wallace had an equally powerful personality, but possessed enough tact to keep in with him. Moreover, all kinds of matters were not receiving the attention they required. Over some issues of joint concern to the company and Asiatic, Cargill had lately had to point out that it was "difficult for Mr. Hamilton and myself, during our rather hurried visit to London, to give the necessary time for discussing this question fully with Mr. Cohen." The solution was to appoint someone who had the board's full confidence, to take full-time control of the London office.

The man chosen was C. K. Finlay. In March 1912 Duncan Garrow succeeded him as manager in Rangoon. Garrow was a cheerful, long-faced man in his early thirties with a nice line in rather good jokes. But he had a deeply sensitive and diffident nature. Cargill, being of much the same disposition and thus rather a bully to those who did not answer back, was unduly overbearing in correspondence with him. At the outset he warned Garrow to take things "reasonably" and leaving the detailed work to his staff so that he could concentrate on "the bigger questions of management."

Finlay was to have a deputy, one currently in Rangoon. He was Robert Irving Watson, another younger son from

Scotland.[1] His father was Thomas Watson, editor and proprietor of the *Dumfries and Galloway Standard*. As a boy he had been a star pupil at Dumfries Academy, eventually as classical dux or top scholar of his year in classics; the confidence he acquired in his own powers was to prove a hallmark of his character. According to an account published in 1922, it was while at school that he heard "the call of the east".[2] That account also attributed to heredity the grasp of detail, breadth of vision, orderly marshalling of facts and arguments, and gift of lucid exposition that made him so successful in his business career. In fact, he would have made a wretched journalist, for in his anxiety to cover every aspect of very intricate questions, his sentences often became excessively involved and not at all clear to his readers.

He had gone out to Burma in 1902, at a time when Finlay Fleming & Co. was greatly understaffed on the oil side. He had proved equal to the steady expansion of work and by 1910 was head of the kerosene department. The same age as Garrow, he looked older because of his baldness. He had shown an unusually thorough understanding of the complexities of kerosene marketing; this was to make him the ideal prospective alter ego of Cohen in the routine wranglings over product agreements. However, Cargill did not care much for his outspokenness, obstinate defence of his views and long-winded style in correspondence. But no one could question the wisdom of testing his gifts on a broader scale than that available in Rangoon.

Watson started in the London office during August 1912 and from the first moment displayed a rare assurance, confidently expressing views to Glasgow on a whole variety of topics. In negotiations with Cohen, a man only a year older than himself but of managing director status, he was at a disadvantage as a mere employee; yet the two men were soon regularly meeting to resolve issues on equal terms. Watson, who was to dominate Burmah Oil's affairs – in fact, *was* Burmah Oil – for the next thirty-five years, had embarked on his spectacularly successful entrepreneurial career.

He at once strengthened the management structure by introducing a statistical section into the London office, taking over geological matters from Glasgow, and reorganizing the correspondence system. He introduced letter books for the weekly Rangoon mail, each separate topic being given its number;

there were also letter books for the correspondence with Asiatic and with the geologists. He still kept the private letter books for more confidential matters. He gave precise instructions on how he expected correspondence to be conducted, for example by discouraging personal letters that touched on business affairs. Not that he had any objection to responsible people writing candidly, but he had to safeguard the fullest flow of information among those entitled to see the various categories of letters. These arrangements may have been to his satisfaction, but were a burden to those in Finlay Fleming & Co., increasing the flow of paper work and also the amount of interference from home that they could well have been spared.

With an almost audible sigh of relief, Cargill eased himself into his new and congenial role above the hurly-burly. Now that Finlay was in charge down south, he had his most intimate friend and confidant in the position where he could be of very great value. If Watson as second-in-command was apt to be a bit abrasive at times, Finlay was there to keep a sharp eye on the correspondence. Watson's *bête noire* was the slackness that he detected in Finlay Fleming & Co. despite some recent attempts by Finlay at reform. While out east in 1913, Cargill was vicariously at the receiving end and privately appealed to Finlay, "For heaven's sake stop Watson's sarcastic and nasty style of writing to Rangoon." His protests soon died down as Watson remorselessly drilled his correspondents there into doing things as he wished.

In May 1913 Finlay was made a local London director, and the following December his nominal chief, M. T. Fleming, died, the last and least memorable of the original executive directors of 1886. Finlay was at once elected to the Glasgow board. He even embarked upon his new post with a knighthood, having become Sir Campbell Kirkman Finlay in the birthday honours of 1912 for his public services with the legislative council and volunteers in Rangoon. In May 1914 he was invited to a levee at St James's Palace; the company ordered an electric brougham and two liveried men to convey him from the Savile Row firm, where he had hired his court dress, to the palace and back again.

Yet something seemed to be wanting in Finlay's position. Sagacious, energetic and highly knowledgeable as he was, he found himself cast as a paper-pusher rather than as the policy-

maker his father had been. Much of his time was taken up with
the affairs of Anglo–Persian, of which he became a director in
1912, but the top decisions there were made by Greenway,
Wallace and Cargill. When important matters affecting Burmah Oil cropped up, Cargill was down by the next train or
briefing Wallace to take charge. At the same time, with lesser
affairs it was always Watson who was in the thick of it. Finlay
must sometimes have wondered if he had been wise to come
home, and if he would see himself becoming redundant in the
long run.

Relations with Cohen were not on the whole to be all that
harmonious over the next few years. Maybe collaboration had
proved mutually so beneficial that both sides could indulge in
the luxury of a good old row from time to time. The first such
row was over a tiny concern, the Twinzas Oil Company, which
Cohen feared might disrupt the product agreements much as
Jamal and Indo–Burma had done. Cargill refused to do anything that might inflate the importance of an insignificant
rival and encourage others to indulge in the now fully established practice of speculating in well purchases. Cohen then
threatened to acquire the concern itself, although he must have
known perfectly well that the authorities in Delhi and Rangoon
would ban such a step. The subsequent fracas went through
the time-honoured phases of voluble indignation on his side
and straight talking from Cargill. "We will keep our eyes
very wide open," Cargill informed Garrow, and stand ready
to "take up the matter strongly with the India Office and the
Admiralty" in order to head off this takeover threat from
Shell.

Often far more trivial issues caused rifts between them. A few
months after the Twinzas Oil episode, Cargill unwittingly
revealed how he was beginning to rely on Watson, writing to
him over a couple of outstanding questions, "I shall be very
interested to see whether your powers of persuasion are sufficient to get Cohen to treat these two items in a broader-
minded spirit than he seems inclined to do." Watson, with
Finlay's full backing, replied, "I am afraid my 'persuasive
powers' are not sufficiently developed to secure a verdict in
such a weak case as Rangoon has given us!" In acknowledg-

ment, Adamson conceded that "Rangoon are a little too grasp-ing," and left him to settle the questions on the best terms possible.

Where Watson felt assured of the company's case he pressed it to the limit. He soon evolved his own routine: Cohen would raise an issue by correspondence, which was settled verbally. Then Cohen or Deterding thought out fresh points and started a new round of debates. Watson's return ploy was to write at inordinate length and demonstrate stage by stage where the new arguments happened to be wrong-headed. Cohen would then come back and perhaps secure a minor concession or two, but more frequently he had to give way.

Another matter that arose at about that time concerned Burmah Oil's tanker fleet. In the summer of 1913, while calculating future transport needs, Watson discovered that the company was short of tanker capacity. As its kerosene trade in India was steadily increasing, an early decision was required. Either the company would have to rely on Asiatic vessels much more than in the past, which it could do under the various product agreements, or it must build itself a new tanker.

When Watson discussed the question with him, Cohen soothingly replied that there was no cause for alarm. He would arrange to provide the necessary tanker tonnage as the com-pany's shipping demands grew, but why not let the Shell group take over the entire Burmah Oil fleet and then arrange to meet all tanker requirements in full? This, commented Watson, was quite a tall order which would hardly commend itself to the company even if it were to involve no financial loss. To Glasgow he observed that Cohen was plainly seeking to collar all the profits on Burmah's tanker operations.

Then fate took a hand when three separate incidents affect-ing the company's tankers followed in quick succession. In Oc-tober 1913 the *Syriam* was in collision and had to be dry-docked for several weeks. In November, the *Twingone* was destroyed by fire while discharging petrol at an Indian port, six lives being lost. A month later the *Singu* had a particularly close shave, again while discharging petrol. Cohen was quick to renew his offer to take over all Burmah Oil's remaining tankers, but Watson used the device of out-Cohening him with an exhaus-tive memorandum that covered every conceivable aspect of the question. Cargill abruptly brought the matter to a close by signing a contract for a replacement tanker.

Meanwhile, Watson had proposed to Cohen the revision of the petrol agreement, pointing out that the demand for that product, together with world prices, had been rising so rapidly that Asiatic should give Burmah Oil a share of these improved rates. As so often, Cohen at first reacted well and agreed to allow the price to rise in line with that of Shell petrol sold to motorists in Britain. Then translating the general principles into cold print took months of acrimonious bargaining, Watson being very much on his mettle as it was the first really major negotiation he had handled.

As late as March 1914 the two men were still at it hammer and tongs. At a meeting, appropriately enough on Friday the 13th, a first-class row erupted between them on how far Burmah Oil should be allowed to supply top-grade petrol over and above the heavy benzine also included in the contract. Cohen hotly accused Watson of scheming to push up deliveries of the top-grade spirit at the expense of the heavier variety, and further inflamed Watson's probably synthetic indignation by hinting that here was a case of "London milk when it was not milk", an obvious reference to the supposed adulteration by dairy-men in the metropolis.

"Never mind London milk," Watson snapped back somewhat humourlessly, "we are discussing benzine." Should that be Cohen's final word, he went on, then they were wasting each other's time. "We would continue the present contract to its bitter and unremunerative end and from and after 1916 we would market our own production." Aware that he had gone too far, Cohen suggested suspending the debate until they had both cooled down. Watson stuck to his guns and insisted that "no number of adjournments would alter our position and we had better just save each other's time and have it settled one way or another now." He then sat Cohen down and took him through the agreement clause by clause to prove that he had no "scheming" in mind.

After some wriggling, Cohen backed down. Yet Watson came away puzzled by the man's uncharacteristic obtuseness throughout the altercation, because hitherto – as he reported to Cargill – negotiations had been conducted on Cohen's side "very pleasantly and, making due and natural allowance for the interests he represents, quite straightforwardly and candidly". He disclaimed any wish to stir up the unfortunate "feeling" that seemed to exist, and continued:

> so long as we get what we want and is our due – getting it nicely if possible, but getting it – and personally I have had nothing to complain [about] so far, I think that on the whole our present relationship is at least the advantageous substance which should not be lightly jeopardized for the, at any rate possibly, less advantageous shadow.

Incidentally, that sentence is an example of the involved Henry Jamesian syntax that emerged from his capacious mind as he dictated.

The whole episode can be regarded as a landmark in Watson's relations with Asiatic. He was clearly warming to Cohen and aimed to discard his predecessors' often irrational suspicions and to work very closely with him, so as to get the best out of him to their mutual advantage. To be sure, that meant barneys galore but no lasting resentment. The respect the two men developed for each other underpinned the company's strategy over the next decade or so.

When battling with Cohen, Watson was all too conscious of the biggest chink in his own armour: the likely future shortages of crude oil in Burma. Finlay, too, had been giving serious thought to the matter ever since his return. The company did have a piece of good fortune in 1912 when its drillers found some fairly prolific wells at Yenangyaung, but early in 1913 those wells tailed off alarmingly. One of Cargill's tasks during his visit to the east in 1912/13 was to inspect the various areas personally and decide on their relative value to the company.

He finally judged that the field with the brightest future was Singu which on its own seemed capable of supplying the refineries' requirements for many years to come. Ideally, however, it needed to be kept in reserve and drilling concentrated in the still productive Yenangyaung field. At the same time, tests outside the main Burmese fields should be kept to the minimum – a decision that drew a sardonic comment from Adamson to Garrow in Rangoon:

> All the oil that has been or ever will be got from these outside fields will never repay the money expended on them. Hope springs eternal in the human breast, however, and I suppose we shall go on

chucking money away punching holes in the ground until there is no room for more holes.

Such worries about long-term reserves underlined some persistent criticisms about the refineries, notably that their fuel consumption was too high and must be reduced somehow. About 7 per cent of throughput went to waste by having to be burnt under the refinery boilers. That was a financial as well as a physical loss: much to Cargill's discomfiture, costs there and in the fields were once again rising. In 1912 Andrew Campbell, the works manager, came home on extended leave. Once at home, he submitted that, for cost-cutting and other reasons, both Burmah Oil and Anglo–Persian needed a department at home to control all technical matters connected with refining.

Campbell's views were echoed by Dr Edeleanu, a Romanian chemist who was approached about that time over a process – later adopted – to remove smoke and smell from kerosene. Edeleanu had been astounded to learn that the company had no "technical bureau", whereas Royal Dutch–Shell had a laboratory at the Hague that was "a model of modern scientific management" where "all processes were investigated in a most thorough manner before they were given to the works." Early in 1914, therefore, Campbell was offered the post of full-time advisory chemist to Burmah Oil and Anglo–Persian.[3] Steeped as he was in the Scottish shale-oil technology that had largely dominated the company's refineries since the 1880s, he now sought ideas from a fresh direction, the continent of Europe.

Despite the advances made by them in drilling and production techniques, United States oil men at that time contributed few refining processes of value to the company. A Californian Trumble still was investigated but rejected because Burmah Oil's own Coffey still seemed better all round. The latter, originally devised in 1866, allowed a number of different products to be distilled off through pipes connected with a tall fractionating column, although heating techniques were not advanced enough to separate out the fractions with any precision. Then Hugh Allan, chief chemist at Syriam, succeeded in adapting the basic process to a fractional atmospheric condenser: that plant he had to enlarge progressively until it achieved the maximum practicable output of the light fractions, which were in greatest demand. Early in 1913, it was experimentally fitted to Bench 1 at Syriam.

For the best part of the next decade there was to be a contest between Allan's process and one invented by Albrecht von Groeling of Vienna. The latter's distillation took place through a dephlegmating column, part of the vapours being condensed and then returned to the column so as to improve the efficiency of fractionation. In 1914 Campbell visited Bucharest to inspect that process, and thought so highly of it that the company ordered the separate preheaters, which reduced fuel consumption during distillation, and also dephlegmators for the Syriam refinery.

Many other aspects of the refineries' operations needed to be looked at as well after Campbell's lengthy reign. The long drawn-out expansion of capacity had really come to an end when the tenth extension to Syriam and a new refinery at nearby Bogyoke – mainly for lubricants – came on stream in 1910. After that date, as total throughput levelled out by 1912, the emphasis was on rationalizing the refinery structure: it was costly and inconvenient to have operations spread over four different sites: Dunneedaw, Syriam, Bogyoke and Yenangyat. Once installed in the London office during 1912, Watson pressed hard for what became known as the Refinery Concentration and Reconstruction Scheme. Yet he could do little as long as the plant and stores department was located in Glasgow; in any case the proposed move on to one refinery site, namely Syriam, would be worth while only when it could be equipped with a reliable and thoroughly up-to-date process.

Watson therefore kept Campbell and the engineer John Gillespie busy investigating every invention to do with refining that might conceivably be of use in Burma. He instructed a firm of patent agents to send the company copies of each oil specification as soon as it appeared. Hamilton does not seem to have objected to this encroachment by the junior man in London on to part of his plant and stores territory.

Another urgent need was to diversify into other parts of the world: something that had not really been feasible since 1904, as so much of the company's financial resources, managerial effort and technical manpower at all levels had been locked up in the Persian operations. By about 1912 the easing of that burden over Persia coincided with the realization that workable quantities of oil in the rest of Burma, despite promising indications here and there, were unlikely to be discovered. That year, therefore, Burmah Oil took a quarter share in some concessions in Trinidad (see Chapter XVIII) and the directors set their geologists on to exploring likely parts of the Indian subcontinent.

The first exploratory well outside Burma, at Sitakund near Chittagong, was unproductive; the next investigation, at Badarpur in Southern Assam[4] where a tea-planter had noted some helpful indications, led to oil being discovered there in 1915. Some all too familiar problems soon followed: poor-quality oil, water seepages and declining yields. Yet the drillers persisted and worked the field to exhaustion, abandoning it in 1933: the total net loss to the company there was said to be about £1,000,000. The company's geologists also put much – ultimately wasted – effort into prospecting in Baluchistan, then in Western India. The energetic Watson was constantly goading them into more determined efforts. Their ultimate failure illustrates that most salutary of lessons in oil affairs: that looking for oil is a very risky business which therefore requires much detailed attention from management. Perhaps as much as a tenth of Watson's time in 1912 to 1924 was devoted to the costly exploration in India and Trinidad, when he had all too much other business to occupy him.

By the middle of 1914 Burmah Oil seemed in better shape than it had been for some time. Its net profit per barrel of throughput, at 18.7p, was the best since 1903, due in part to the many reforms of recent years. A major factor was having the London office in strong and capable hands. There had been a change in Rangoon too. Duncan Garrow had recently come home and joined Anglo–Persian, which for him was a far more congenial job; he soon became a director. As general manager in Rangoon he was succeeded by Adam Ritchie, a 33-year-old Cambridge graduate and chartered accountant. Ritchie was more in the Finlay tradition, being a Scot (Garrow was English), an active rather than contemplative type and never shy about arguing his case with the people at home. His willingness to argue made for a more equal relationship between him and Cargill.

That June, the London Office moved to more extensive premises at Gresham House in Broad Street, where Anglo-

Persian was already installed on a different floor. By then the government acquisition of a majority shareholding had relieved the company of long-term financial anxiety over Anglo–Persian, and the Mesopotamian question was as far advanced as those involved could reasonably expect. A less hectic period of consolidation was therefore on the cards.

Finlay was to oversee all the London office affairs while Watson went on holiday for the whole of August. Greenway, having endangered his health by all his exertions of the previous six months or so, also began his holiday at the end of July, knowing that the legislation was proceeding smoothly through parliament. As Wallace planned to be away as well, Nichols, lately appointed assistant managing director, would be in charge at Anglo–Persian. Then the unthinkable happened. Early that August, Britain was plunged into war.

Notes

1 See obituary in *Dumfries and Galloway Standard,* 31 January 1948.

2 "Mr Robert I. Watson", in J. Reid, *Some Dumfries and Galloway Men* (1922), pp. 250–6.

3 The history of the company's research department (from 1916 in Putney and after 1921 at Fairlawn, Honor Oak Park, south London) is given in *Burmah International,* 1, December 1969, p. 30. For A. Campbell's description of the refinery process at about that time see his paper, "Petroleum Refining", *Journal of Institution of Petroleum Technologists,* 2, 1916, pp. 274 ff. and his book *Petroleum Refining* (1918).

4 For Badarpur see *Burmah Group Magazine,* 10, 1967, pp. 33–5.

CHAPTER XV

The First World War
1914–18

For Burmah Oil there were no heroics in the First World War. No vast operations had to be improvised against time, regardless of cost; nor was there the excitement of organizing a great flow of new products to help encompass the destruction of the enemy. In contrast with many other British firms, therefore, the company seemed to resemble W. S. Gilbert's House of Peers which, in an earlier conflict, did nothing in particular, and did it very well.

Yet this was the first major war in which petroleum products were to be decisive: as Lord Curzon, a member of the war cabinet from 1915 onwards, put it in November 1918, "the allied cause floated to victory upon a wave of oil."[1] That wave in part had its source in Burma. The company supplied such fuel oil as the Royal Navy needed in the east, and a good deal of petrol for wartime needs. For the rest, it kept the home lamps burning throughout the Indian empire: no inconsiderable feat in the later years of the war when kerosene supplies by Asiatic and Standard Oil were cut back or halted altogether.

All in all, it was a fallow period for the company, which had to do the best it could with techniques, such as drilling and refining methods, that were largely outdated. Whether the inside directors, all of whom met in Glasgow on 5 August 1914, had the will to force change through is doubtful; but in any case, by armistice day 1918 only Cargill among them remained fully active in the company. Hamilton, Adamson, Finlay, Wallace and D'Arcy were either dead, incapacitated or occupied elsewhere. The man who in 1914 was only second-in-command in

London, ascended step-by-step to a dominant position in the company, being made a Glasgow director in July 1918. That was R. I. Watson, and if the conduct of wartime activities were to be specifically associated with anyone in Burmah Oil, it was with him.

He did in fact apply to join the army, but was told to stay at his nationally important job. Instead it was the 38-year-old Finlay who, as a former lieutenant-colonel of the Rangoon Volunteer Rifles, secured a senior appointment in the field. He was soon spending week-ends on his military duties, helping to raise the 13th battalion of the Northumberland Fusiliers, of which he took command in December. Then before he could go overseas he was found to be suffering from angina pectoris. He was invalided out and at once returned to the London office, only to find that things had changed there even in the few weeks he had been away.

R. I. Watson was dashing off letters to government departments and rival companies as regularly as Kirkman Finlay had done in days gone by. The younger Finlay therefore had all too little to do, and early in 1916 joined the Ministry of Munitions. Later on he became – of all things – head of the oil seeds branch of the Ministry of Food, where he stayed until 1920.

At the outset of the war, the authorities were especially anxious about fuel oil, provisions for which were included in the government's "War Book", drawn up by a sub-committee of the Committee of Imperial Defence.[2] When Russia and Germany mobilized on 31 July 1914, Admiral Slade, then on special service in the Admiralty, invoked the war clause in the 1905 fuel oil agreement with Burmah Oil requiring the company to produce up to 100,000 tons a year. Finlay at once pointed out the losses of other products and of earnings that would follow; gradually Slade scaled down the annual requirements to no more than 10,000 tons. Even then the navy took only such deliveries as it currently needed, for with Russia an ally and Japan a friendly neutral, no theatre of war was expected in the east. The total of fuel oil actually taken was not considerable: just over 11,000 tons a year, or roughly half as much again on average as in the six years 1908–13.

Only once did hostilities come at all close to the Indian empire. The German cruiser *Emden* was reported to be on a raiding expedition in the Bay of Bengal, and late in September bombarded Burmah Oil's storage tanks for kerosene and fuel oil at Madras, scoring direct hits and setting two tanks on fire. Cargill raised a rare laugh at the next annual meeting when he declared that in due course he would be rendering a claim against the German government for the £7,800 worth of damage done.[3] Cohen at once arranged to make up the company's losses of products until the damage was repaired. The *Emden*'s activities caused great disruption to tanker movements until she was sunk in November 1914.

The company's task of turning out the maximum practicable amount of kerosene was important enough: as Finlay Fleming & Co. later expressed it, the people of India "would otherwise have been left in darkness". But as the war dragged on and became more mechanized, supplies of petrol and benzine became a top priority. The company stepped up production of both by nearly 30 per cent, sending three-quarters to Europe and well over half of the remainder to various fronts, notably against the Germans in East Africa and against the Turks in Mesopotamia.

Yet only 4 per cent more crude oil passed through the refineries in 1918 than in 1914. To squeeze out the best mix of products, the works staff had to use the plant well beyond prudent limits and to manufacture their own replacement parts where these could no longer be imported from the west. The pipeline, too, was transporting half as much again as it had originally been designed to take, with consequent wear and tear in the pumping stations.

A fresh worry about inadequate crude supplies generally had afflicted the directors just before the outbreak of war. The flow of crude to the refineries had lately declined so fast as almost to exhaust buffer stocks. Ritchie hastily arranged to drill every available site in Yenangyaung, leaving Singu as the company's main stand-by area. Cargill stifled his earlier anxieties about ever-rising costs in the fields and agreed to the recruitment of eight extra drillers, bringing the numbers up to ninety-nine in all; he did not blench even when Hamilton had to execute a "formidable" order for drilling materials and stores.

The extra drillers and equipment arrived just in time to prevent a further fall in output as wartime demands were building up. Then for the first time the insistent voice of Watson was to be heard on policy matters. As the one having to keep his end up with Cohen over the product agreements, he had for two years been urging on Rangoon a far more energetic drilling policy. He now put it to Cargill that even more drillers were needed, if only because rival companies there had much greater numbers in relation to their output. Nor could he see any point in keeping Singu as a reserve. "This tenderness to Singu is, I suggest, misplaced, or 'reserve' does not carry the common meaning of the word. If the 'reserve' does not exist for just such a situation as the present, then I really cannot imagine what it exists for."

Cargill did not rebuke Watson for speaking out of turn. Instead he stood fast over the drillers and Singu reserves but discussed with him how far management on the spot might have been to blame for recent shortages. The fields superintendents still seemed to have "long-chair" attitudes to their work: one had recently been warned in no uncertain terms "to go round each area to see things for himself, and not sit still in his house merely issuing instructions". But the fields manager, C. B. Jacobs, was plainly nearing the end of his usefulness. The unsentimental Watson was only too glad to see the back of him when late in 1914 he expressed a wish to retire. He was succeeded by his assistant, John Seiple, who had been in Burma since 1897.

The company at once sought to discover what exactly had gone wrong in the fields, sending out W. L. Mackenzie, an engineer and director of its equipment-ordering subsidiary, the Oil Well Engineering Company. For the first time an outside investigation was being made into an aspect of the company's work. Mackenzie spent two months in Burma at the end of 1915 and submitted his report the following February. That report – no longer extant – Cargill found "unpleasant reading". Above all, it disclosed shocking extravagance, notably in the wastage of stores. The remedial measures of 1911, described in Chapter XI, had clearly proved inadequate. Cargill at once called Ritchie home so that they could go into the matter fully. Meanwhile, with soaring costs of all drilling materials and of sea transport, he had the drilling programme cut back where possible, especially on deep wells that yielded only meagre amounts of crude.

At a subsequent meeting in London, Watson and Ritchie

were startled to learn that the Glasgow directors had accepted Mackenzie's recommendation that Seiple should take full responsibility for all fields matters, including the drilling staff. The fields agent would therefore confine himself to office matters and relations with the Twinzas. Seiple had in fact issued an ultimatum that he must be fields manager in reality as well as name: that meant sacrificing all the progress made by successive fields agents since 1905. Uncontrolled and costly Yankee management seemed to be back with a vengeance. Watson was also unhappy that Seiple would once again be landed with a mass of correspondence as well as detailed consultations with geologists, which the fields agent had recently taken over. He was overruled, Cargill sharply reprimanding him for criticizing decisions already taken and forbidding him to mention his disagreement to Rangoon.

Mackenzie's report did lead to some practical economy measures, such as stamping metal parts to deter pilfering, and using employees rather than contractors for rig-building. As Seiple grew into the job, his able conduct helped to allay Watson's fears: he managed with fewer drillers and ancillary labour in the fields and was known to "stand no nonsense" with those under him. He seemed to be coping adequately with the office work and collaborated well with the fields agent. Fields expenditure in 1917 and 1918 was therefore only a little above the level of 1913–14, at a time when prices generally were doubling. To be sure, materials were by then so scarce that economy was forced on everyone, with old equipment having to be re-used and repaired.

Seiple's tough attitude, and the decision to keep drilling down to the bare minimum required to meet refinery needs, proved their worth in April 1916, when sixty-eight drillers went on strike for higher allowances. The Rangoon office at once dismissed them and hustled them out of Burma. He undertook to manage with the fifty-five who remained at work; they were given higher wages and improved fringe benefits in return for promises of better timekeeping.

Overhauling fields administration was admittedly of value, but one certain source of economies would be the electrification of the fields. It was totally inefficient to have separate steam boilers to power the drilling and pumping engines, which by

this time burnt up 10 per cent or more of the crude raised. As
wells went deeper, so crude consumption on the spot rose. Yet
as early as 1908 Redwood had advised the directors that fields
in the United States and Russia had shown how electric power
could prove safer, cleaner and more convenient to work. Indeed, electrification was successfully going ahead in Persia.

Any serious moves by the company had hitherto been ruled
out by Jacobs's unwavering opposition. Seiple, as his successor,
proved far more amenable and co-operated with some trials at
Singu, using experimental equipment shipped out by the British Thomson-Houston Company of Rugby, a subsidiary of the
(American) General Electric Company. The trials were most
encouraging, and Cargill and Hamilton decided to put the
plant and its installation out to tender.

Watson at once raised objections. No firm in Britain had real
experience of electric drilling and pumping, he pointed out, so
that the company might in due course find itself saddled with
quite unsuitable equipment. A consultant from British Thomson-Houston was already engaged on the Anglo–Persian electrification scheme; should he not be asked to draw up a
comparable scheme for Burma? Watson made the proposal
with some diffidence, for Cargill could be touchy about
attempts to interfere with Glasgow's responsibility for plant
and stores. In fact, it was accepted without demur, and Watson
was told to get on with the whole scheme. Once the detailed
specifications had been worked out, it was put out to tender in
1918. Not for the first time, Watson's candid observations had
led to a major issue being handed over to him.

Of all the anxieties which the war caused the directors, none
was more persistent than tanker capacity, since by then about
85 per cent of kerosene and other products found a market
outside Burma. Although the company had six tankers in
service or under construction, the *Twingone*'s replacement was
making little progress in a British shipyard as it was of low
priority. A tanker had been secured on temporary charter, but
that charter was unlikely to be renewed in wartime conditions.

Mercifully, the company was safeguarded by the product
agreements that required Shell to undertake transport if necessary. Shell, which owned a world-wide tanker fleet, had already
offered all its tonnage to the British government at very moder-

ate charter rates, and many were already on war service. Cohen
pledged himself to keep Burmah Oil fully supplied with tankers
for the duration, but expected the company to use its own fleet
as economically as possible. Even Cargill, already worried to
death by the war and at the best of times never one for looking
unnecessarily on the bright side, regarded that undertaking as
"highly satisfactory", the more so as his own tankers were very
vulnerable to official requisitioning at short notice.

Then a tricky point of conflict arose out of the desperate
shortage in the Rangoon refineries of fuel for their operations.
Supplies of their own *astatki* or residue – used for that purpose –
were inadequate by 10–15 per cent, which had to be made up by
using crude oil that could be ill afforded under wartime pressures of demand; but Abadan had plentiful residue that would
be ideal for refinery fuel, and the obvious solution was to ship it
to Rangoon. The company arranged to do so, much to the
chagrin of Cohen, who resented having to divert scarce tankers
for that purpose. Despite many protestations, however, he did
not let the company down.

Watson was also collaborating advantageously with Cohen
over the establishment of joint bulk petrol installations at the
four major Indian ports. These would avoid unnecessary du-
plication and ensure that neither party stole a march on the
other. They set up a jointly owned registered company, the
Tank Storage (India) Ltd. Ultimate authority was in London,
but day-to-day matters rested with their respective general
managers in Calcutta. It thus differed in kind from the wax
pool, which had neither a formal company structure nor any
fixed assets comparable with the installations. The arrangement was to prove a historic one, as it was to be used in 1928 as a
nucleus of the Burmah–Shell Oil Storage and Distributing
Company of India Ltd (see Chapter XVIII).

Overlapping those discussions were longer-running differences of view over the company's maximum price policy for
kerosene. The Shell group detested that policy and in April
1915 Cohen proposed to Watson that as all costs, including
transport costs, had risen so sharply since 1914, official permission should be sought to raise the maximum price. Watson, just
off to take the holiday he had forfeited when war broke out,
penned a quick letter to Hubert Heath Eves, general manager

in Calcutta, asking him to sound out the government of India.

Greenway, as a London director, saw a copy of the letter and at once submitted to Cargill the inadvisability of even hinting at such a thing to government. Should they not let sleeping dogs lie, since the consumer would otherwise have to bear an extra burden on top of higher prices for foodstuffs and other necessities? Moreover, the company was vulnerable to criticism, or even reprisals, as it was making good profits and paying up to 30 per cent dividends. Greenway had a point and Cargill, as the only begetter of the maximum price policy, immediately cabled Eves to disregard Watson's letter when it arrived. Watson, when later carpeted for his transgression, made a spirited defence but was the more careful afterwards not to step beyond his authority.

Cargill told Cohen there was nothing doing, but at his annual meeting shortly afterwards did stress that the company had not bound itself over prices for all time. In February 1916 he had second thoughts and requested Watson's advice: perhaps the first time he had consulted him as an equal on a major issue. Watson, in a closely reasoned reply, said he saw no good reason for seeking to raise the maximum price.

As he had anticipated, the government of India made a strong plea for price levels to be maintained for as long as possible and asked provincial governments to impose ceilings on retail kerosene prices so as to deter profiteering. Matters rested there until Asiatic, having blown hot and cold during the first half of 1916 over supplies from Borneo, finally resolved to stop exporting all kerosene to India as from the beginning of 1917. The decision was mainly due to the lack of tanker capacity. As a consequence, total sales in India for 1917/18 were a quarter down on the record year 1914/15. Since demand was as high as ever, profiteering among dealers became widespread as the official controls proved largely ineffective.

Where the shoe pinched most acutely was over tins. All tinplate had to be imported, and British firms were quoting such lengthy delivery dates that the company was forced to order from American suppliers. Then a deputation from the Welsh tinplate industry complained to E. G. Pretyman, who had recently joined Asquith's coalition government as parliamentary secretary to the Board of Trade, about the company placing orders overseas. Although the matter was smoothed over, tinplate supplies were to remain very tight for the rest of the year.

In 1916, the celebrated Parsee enterprise, the Tata Iron & Steel Company Ltd, offered to manufacture tinplate for Burmah Oil at its plant at Jamshedpur, 150 miles west of Calcutta. At first Watson was uncharacteristically hesitant. Except in such closely related areas as oil-well equipment and bauxite, the company had always refused to join in any outside venture. But the shortage and mounting cost of tinplate made Watson decide, no doubt against Cargill's better judgment, to follow up the Tata offer. His investigations included having plate made up experimentally in Britain and the United States from Tata steel; those won him over, and in March 1918 he signed heads of agreement with Tata.

Meanwhile, the company economized in tins by selling as much kerosene as possible in bulk and by getting the government of India to agree to increased prices for tinned kerosene. Regrettably, it was the consumers in Burma who suffered, as only in India could the railways transport kerosene in bulk. Ritchie, when reporting to the directors the official consent, fervently hoped that the company would never again tie its hands by making promises to government over prices, since the future was always unforeseeable. Watson felt he could indulge in some pawky humour at Ritchie's expense:

> Unfortunately the tactics of the ostrich, however temporarily satisfactory to the bird itself, in no way change the landscape, and our landscape is that this guarantee has been in existence since 1905 and has been made a great deal of both by ourselves and by the Indian and home governments. No burying of our heads in the sand can change this, and it seems to us that we must recognize the position as it exists and not as we would like it to be.

Cargill's reply to Ritchie was non-committal. The end of the war was still a long way off, he felt, and therefore the company's post-war strategy could not be laid down until conditions could be forecast more precisely. Not that those conditions were likely to bring comfort to Burmah Oil, he added with his customary good cheer; he could anticipate only a "battle of giants", including Anglo–Persian, in which case price cutting would make academic the question of holding to maximum prices in India.

The most noteworthy of the old directors to go was C. W. Wallace who, in August 1915, suffered yet another "attack", whereupon his doctors compelled him to go into full retirement. He was succeeded on the Glasgow board by Ashton, and died a year later at the age of 60. Had he been a more robust man, he would undoubtedly have become managing director in the London office after Kirkman Finlay's death in 1903, and five years later would perhaps have had the choice of staying with the company or becoming chairman of Anglo–Persian. He knew his own mind and, if he lacked the sensitivity of Finlay father and son or the sheer grasp of detail shown by Watson, he was never short of stimulating ideas, some culled from most unusual sources.

In middle life he was greatly taken with a book of fictional letters supposedly written by a self-made American merchant to his son, but actually by G. H. Lorimer, editor of the *Saturday Evening Post*.[4] Those letters, spiced with wry humour and homespun wisdom, made the book a best-seller on both sides of the Atlantic. Some of Wallace's own letters which, if edited, would make up quite a useful little tract on business principles, and memoranda which less than tactfully dissected his colleagues' shortcomings, suggest that he would not have been averse to being called, after the title of Lorimer's sequel, Old Gorgon Wallace. It was from that sequel that he derived his most striking text, which he had framed and hung on the wall of Shaw Wallace & Co's Calcutta office:

> It has been my experience that when an office begins to look like a family tree you'll find worms tucked away snug and cheerful in most of the apples. A fellow with an office full of relations is like a sow with a litter of pigs – apt to get a little thin and peaked as others fatten up.

He therefore decreed that no one could join Shaw Wallace & Co. who was related, within the limit of second cousin, to anyone already in the firm. That would ensure absolutely equal opportunities for everyone; whether he ever discussed those ideas with John Cargill is not known, for neither John nor his brother David Cargill nor their solicitor cousin James Thompson would have relished being likened to grubs in the Burmah Oil apple. Wallace's will pursued this theme even further:

Subject to the possessor's right during his lifetime to the enjoyment thereof and to the making of adequate provision for his children until they are old enough to provide for themselves, all possessions great or small being acquired from or through the people, as mine were, should return to the people.

He therefore bequeathed his residual estate to the treasuries of Britain and of India. A clause – never invoked – stated that if his sons were to become baronets or above, they should inherit enough capital to keep them in an appropriate style of life. As he was highly regarded in Whitehall, he may at one stage have been offered a title, but he never accepted one. His will was not finally settled until fifty-five years later, when £1,000,000 was paid over to the British exchequer in 1977. He could not have foreseen that by then, a million pounds would have kept the government machine going for no more than twelve minutes.

Some months after Wallace's demise, W. K. D'Arcy died in 1917, aged 67. According to Cargill, he was "another of the many victims to Christian Science of which he and his people were such staunch supporters." Cargill recalled him affectionately as a "large-hearted, genial man". Although a cypher in Burmah Oil affairs and latterly in those of Anglo–Persian as well, he would always be remembered as the man who, by his courage in acquiring the concessions in Persia, had changed the face of the middle east and hence, utlimately, of the industrial world.

To round off the roll of the senior directors, R. W. Adamson retired in June 1918 at the age of 62, dying of cirrhosis of the liver three years later. He was the one who stayed put in Glasgow, commenting pithily on all topics referred to him, while Cargill and Hamilton were off in London or elsewhere for meetings. James Hamilton's health collapsed in the spring of 1919 and he resigned the managership of Burmah Oil. Six months later he died of arteriosclerosis: he was 60. Apart from Cargill father and son and Watson, the executive directors of the company were poor insurance risks.

During the war, Burmah Oil and Anglo–Persian did not enjoy the close relationship with each other that might have been expected. By 1916–17 the latter company had begun to acquire an identity of its own. In that financial year to March 1917 it had at last earned enough to pay its first ordinary dividends, and was able to convert into debentures the £200,000 odd that Burmah Oil had advanced to meet earlier preference dividend payments. Its accounting and plant and stores departments were transferred from Glasgow and, once the two companies' clerical staffs had been divided, the whole of Anglo–Persian's British operations were together in London. Towards the end of 1917 they came under one roof in Britannic House, Great Winchester Street; there Burmah Oil joined it on the second floor in March 1918.

Perhaps there was a feeling in the young company that its parent was retarding rather than assisting its development. Two or three of its most influential directors were seasoned Burmah Oil men who were wary of the increasingly self-confident Greenway. Although they knew all the important happenings in Anglo–Persian, they rather meanly gave him no more than a limited foothold in their own company and severely restricted the range of Burmah Oil documents he could see as London director. At the same time, he was fearfully overworked as managing director as well as chairman of Anglo–Persian. Wallace seems to have been of little use in the year before his retirement in 1915, and was not replaced for some time as vice-chairman. The pair who ran the day-to-day operations of this rapidly growing company, Duncan Garrow and H. E. Nichols, were not much involved in top policy-making.

Then towards the end of 1916, Slade was appointed vice-chairman when he was ousted from the Admiralty in the Whitehall shake-up after Lloyd George's appointment as prime minister. The Burmah Oil people had had their doubts over Slade as early as 1914 when he had sought to foist a naval engineer-captain on Anglo–Persian's London office as a kind of progress chaser for plant ordered for Persia. Cargill had swiftly nipped in the bud that attempt by a government director to interfere in matters outside his jurisdiction. Nor could it have helped relations much when Slade was found to be plotting to unseat Greenway and have himself made chairman; the coup failed, and in January 1918 he was stripped of all executive responsibilities.[5] To compound the difficulties between the two companies, the merger scheme (see Chapter XVI) seemed calculated to bring unforeseen changes to both and was already causing discord between Cargill and Greenway.

Early in 1918 the Burmah Oil influence received a significant fillip when Finlay resigned from the Anglo–Persian board and Watson was elected in his place. Watson had been indirectly connected as the representative of what was then its marketing subsidiary, the British Petroleum Company, on the Petroleum Pool Board which controlled supplies of refined products in Britain. On Cargill's instructions he had been reading all the Anglo–Persian correspondence coming into Burmah Oil's London office, but had rather ostentatiously refrained from making any comments. When at last he felt free to discuss its problems with Cargill, as he almost immediately did, he could therefore do so from a position of considerable knowledge. The use Anglo–Persian made of Watson after 1918 will be discussed in Chapter XVIII.

As a newcomer, Watson did not automatically go along with earlier fears in Burmah Oil of the powerful "octopus", Standard Oil. During the second half of 1917, following America's entry into the war, Cohen visited the United States as one of Lord Northcliffe's economic mission, and helped to set up the Inter-Allied Petroleum Council to regulate and pool the supply and transportation of oil products for the war. As he explained to Watson on his return, the success of the mission had been hampered by Standard Oil's resolve to protect its own interests by imposing "unfair and unequal sacrifice" on its British counterparts: something it allegedly had the power to do because of its strong influence over the American administration. At first Watson thought he was putting it on a bit; yet when an oil delegation arrived from the United States and made a number of harsh proposals to the European allies, he wondered if Cohen was not right after all.

Standard Oil had already reduced its own kerosene deliveries to India by over half and diverted them further east to China and Japan, hitherto regarded as British markets. Then in December (Sir) Basil Kemball-Cook, virtual controller of all British sea transport at the Ministry of Shipping, passed on a suggestion from higher authority that as Rangoon's heavy petrol – for motor vehicles – was not a top-priority war requirement, Burmah Oil should cease to ship it to Europe and either store it or burn it off. As neither Anglo–Persian nor Asiatic had been asked to do the same, Watson could only assume that the company had been specially singled out "as the line of least resistance".

Kemball-Cook explained that the Americans were seeking to economize in tankers by more rational planning of product deliveries; understandably so, as the whole of Burmah Oil's petrol shipments were currently being undertaken in tankers requisitioned from Standard Oil. Since Europe and the Mediterranean were nearer to the United States than to Rangoon, Standard Oil had offered to supply all requirements for the European theatre of war from its own stocks. Watson's retort was that American petrol was of unacceptably low quality; in any case the United States, now that it was in the war, would require for its own needs all the petrol it could produce.

Whether as a result of his protests or not, the suggestion was dropped. Until the end of the war, however, tanker tonnage became more and more hard-pressed and Rangoon was forced to hold disturbingly high stocks of petrol. Watson still suspected that the machinations of Standard Oil were at work, and as late as October 1918 was complaining vigorously to the Petroleum Executive about instances where the United States authorities were putting "tonnage pressure" on Britain in a way that could only harm Burmah Oil. The war ended before such complaints could be investigated.

Armistice day, 11 November 1918, was celebrated very differently in different parts of the company. Ritchie reported the wild and spontaneous celebrations that broke out among the Burmese as soon as the news arrived by cable in Rangoon, with impromptu processions of all races and creeds around the town. As he wrote home:

> When I arrived in office on the morning of the 12th, I was met by a native clerical staff cheering at the tops of their voices, so the only things to do was give them Rs.450 [then about £45] amongst the lot to enjoy themselves with.

Official victory celebrations took place at the end of the month, with a night-time military tattoo and illuminations on Rangoon's maidan, or principal park. Ritchie estimated that well over 100,000, or a third of the capital's population, were present.

In the London office, Watson must have given the staff a holiday, but there was none for himself or his secretary. He had of late complained that he was spending more time out of the office than in, because of "meetings here, there and everywhere lasting nearly all day", and had viewed his recent promotion to the Anglo–Persian board as simply "a lot more encroachment on an already overburdened time". So he used an unencumbered day – Deterding had already sent home Asiatic's staff – to get through a large number of documents and deal with a whole backlog of correspondence.

Unlike Cargill, who on the 12th expressed heartfelt relief that it was all over, Watson never once mentioned the momentous event in his letters of that day. Nor would it have struck him to recall how month by month and year by year he had made himself ever more indispensable to the company: not out of overweening ambition, but merely in response to business demands. He was too busy looking ahead, and indeed, within three months had produced his blueprint for the company's post-war role.

Notes

1 This often misquoted phrase was in *The Times*, 22 November 1918.
2 The "War Book", produced by the Committee for the Co-ordination of Departmental Action on the Outbreak of War (under the Committee of Imperial Defence), dealt with the departmental measures to be undertaken when moving to a state of war and became active in 1914: see PRO CAB 15/5.
3 AGM 1915, *Glasgow Herald*, 17 June 1915.
4 G. H. Lorimer's two books were *Letters from a Self-Made Merchant to His Son* (1903) and *Old Gorgon Graham* (1905). The "worms" quotation was from the latter book, p. 4. The limits on consanguinity in the partnership and Wallace's will are mentioned in Townend, *History of Shaw Wallace*, pp. 23–4. For settlement of will see *Daily Telegraph*, 25 May 1977.
5 Ferrier, "Early Management Organization of British Petroleum", (see reference 11 in Chapter XII), p. 135.

CHAPTER XVI

The Merger that Never Was: (1) The "Imperial Question" 1915–19[1]

By 1914 Burmah Oil was no longer the insignificant pawn it had once been in oil affairs. Although that year it produced only a little over 1 per cent of world output, it now had an importance as the largest supplier of kerosene to India, itself the most extensive single market outside Europe and North America. As a result of the Admiralty fuel oil agreement, it could count on a measure of protection from the British government, if less certainly from the local administrations in Delhi and Rangoon. Its relationship with Whitehall had been further strengthened when, quite unexpectedly, it became a partner of the government in the ownership of Anglo–Persian.

Although Burmah Oil itself had a reasonably good relationship with the Shell group, ever since 1909 Deterding, and through him Cohen, had striven to blunt the impact of Anglo–Persian. Their not very successful efforts in Egypt and Mesopotamia have already been mentioned, but they had had a quite unforeseen bonus when Anglo–Persian asked for and was granted product agreements with Asiatic, which were to run up to 1922.

Yet any attempts by Asiatic to use these agreements as a lever for securing control of Anglo–Persian signally failed. Greenway had skilfully used Cohen's alleged further designs on his company to induce the British government to buy a controlling interest. Thereupon the Admiralty, alarmed at the stringent terms of the Asiatic–Anglo–Persian agreements, looked hard for means of breaking them either by challenging them in the courts or by reorganizing the Anglo–Persian holdings so as to make those agreements null and void. Naval circles were already prejudiced against the Shell directors. For instance Black, who was ultimately involved in oil procurement as Director of Navy Contracts, regarded them as "good servants but bad masters"; in other words "excellent contractors whose usefulness we appreciate, even if rather adept at doing a one-sided bargain'. Not that the more kindred spirits from Burmah Oil evaded Black's strictures:[2]

> The Shell group is well aware of the vulnerable spots in the Burmah Oil Company's armour. Whenever they wish to extract some concession from the Burmah Oil Company or the Anglo–Persian Oil Company, they threaten a war of rates in the Indian market. Mr. Cargill, the chairman of the Burmah Oil Company, is peculiarly susceptible to this kind of pressure. His unfortunate weakness in this respect led to his saddling the Anglo–Persian Oil Company with severely restrictive marketing agreements with Shell.

Maybe Cargill, for all his lugubriousness, was not quite so lacking in guts as Black suggested. The maximum price for kerosene in India still held, and Black never grasped the almost insuperable difficulties confronting Anglo–Persian when the product agreements had had to be concluded. In any case, official hopes of exploiting legal loopholes in the agreements faded when in June 1915 counsel's opinion advised the Admiralty that the terms were legally watertight.[3]

A few weeks later, at the end of June, Cohen unfolded to Cargill and Watson a dramatic proposal, involving an outright merger between Burmah Oil and Royal Dutch–Shell. He claimed that such a merger would secure British control over the whole group, then 60 per cent Dutch-owned and 40 per cent British. An exchange of shares would make Burmah Oil a wholly-owned subsidiary of the group, but it would maintain its separate identity as a trading unit. That would provide a definite if narrow British shareholding majority, which would be safeguarded by banning share sales to foreigners, and by giving Burmah Oil and (British) Shell enough seats on the board to outvote the Dutch directors.

Cargill's instant reaction to any proposal of substance was to turn it down. As he informed Ritchie in the strictest confidence, he could see neither Whitehall nor Delhi for a moment entertaining Cohen's scheme. To Cohen he made it clear from the outset that he and his board would refuse to consider the proposal formally until invited to do so by government, thus putting the onus on the authorities to accept it first. Meanwhile, he unenthusiastically passed it on to Slade, by then personal adviser to A. J. Balfour, First Lord of the Admiralty.

Cargill conceded it had attractions for his own shareholders, who would gain a direct stake in the Shell group's expanding world-wide business and no longer be limited to the wasting Burmese fields. He reserved the sting for his final words: "As after all the directors of the Burmah Oil Company are really the trustees for the shareholders, it is their duty to give it their most serious consideration before absolutely refusing to consider it."

Slade's comments, made to his superiors in the Admiralty, were predictably hostile as well as wide of the mark. He refused to believe that Shell's directors were sincere in seeking British control. Moreover, incorrectly assuming that Cargill was all for the scheme, and wishing to protect Anglo–Persian – of which he himself was a director – from even indirect shareholding links with Shell, he minuted, "Of course it is open to the government as the controlling power in the Anglo–Persian Oil Company to remove the whole of the Burmah directors [from the Anglo–Persian board] and replace them by their own nominees – not a policy I would recommend except as a last resource. It would be difficult to find suitable men and the government does not wish to take over the commercial working of the company."

Slade's aim was not merely to defeat the Shell scheme but also to promote an alternative one. Whether Cohen knew it or not, Lord Cowdray, founder of the Mexican Eagle Oil Company Ltd which had large oil interests in Mexico, had in December 1913 offered to keep the Admiralty supplied with fuel oil in return for an investment by Whitehall of £5,000,000 in his company. The offer was made to Lloyd George, then Chancellor of the Exchequer, who at that moment had a better bargain over Persian oil for less than half that sum and therefore declined. Cowdray had later approached Anglo-Persian with proposals for a joint marketing company in Britain, which he hoped Burmah Oil would join as well. In

1915 the project seems to have blossomed into specific merger proposals; yet Black saw little merit in them. To him, Cowdray was a strong-willed man who would dominate the combine and ignore any directive of the government despite its majority shareholding in Anglo–Persian. Besides, he continued, "the vacillating policy of the Burmah Oil directors will still be a factor."[4]

However, it was Cohen's scheme that all interested departments in Whitehall had to consider, and Black therefore invited comments from the India Office, informing the Treasury, Foreign Office and Board of Trade as well. In mid-September the India Office replied that the government of India found the scheme totally unacceptable: the underlying identity of views had already been anticipated by Ritchie on the grounds that Delhi's policy was "nowadays so largely the policy of Downing Street". That rebuff for the moment halted the scheme's progress.

Cohen, however, tried a new ploy. He called on Burmah Oil to reduce drastically its kerosene deliveries to India on the grounds that Shell needed an outlet for their relatively low-grade Borneo kerosene which was being refined as a by-product of the petroleum-based explosive, toluol, claimed by Cohen to be essential to the war-effort.

Resisting this demand, Cargill, Finlay and Watson discovered at a meeting at the Admiralty that Borneo toluol was not as vital an issue as Cohen had implied. Yet Burmah Oil could leave nothing to chance. Hamilton in Glasgow and Finlay, Watson and Andrew Campbell in London were making "furious calculations" on the total compensation they would require if they had to give up part of the Indian market. No sooner had they agreed on their bill than Cohen decided to make alternative arrangements. The toluol was instead produced in Egypt where the kerosene could be readily sold, and later on in England.[5]

Cargill's alleged spinelessness, that had so exercised Black and others, was put to the test in October 1915 when Cohen made a serious complaint to him about Anglo–Persian. The dispute at issue was a minor one, but one of many, and Cohen claimed to be powerless to avoid constant friction. He was no doubt within his rights to raise a specific issue with Cargill as an important board member of Anglo–Persian. It was quite another thing to make a more general attack:

> Some day the position between the Anglo–Persian and ourselves must be cleared up. . . . It has always seemed to me an impenetrable mystery that our relations with the Burmah Company should be so pleasant and mutually profitable, but that with the Anglo–Persian Company, in which you hold such an overwhelming interest [*sic*] we should constantly find the opposite to be the case.

When Cargill tried to play down the whole matter, Cohen insisted on pursuing it by making two charges. First, Cargill was allowing himself to be swayed by the misrepresentations of certain other Anglo–Persian directors: clearly Greenway. Cargill demolished that one without ceremony, seeing it as not only an insult to his own intelligence but also an unacceptable slur on his colleagues. Cohen's second assertion was that he always found it so easy to adjust differences with Burmah Oil, in contrast with Anglo–Persian. Cargill retorted that on many occasions agreements between Burmah Oil and Shell had been reached only after long and acrimonious discussions and correspondence, many of these arrangements giving Burmah Oil less than fair treatment.

That quarrel was halted by the fortuitous publication of an article by Sir Marcus Samuel in a maritime journal berating the British government for its Persian oil deal[6]. Cargill took the heaven-sent opportunity to extract from Cohen a declaration that his board dissented from Samuel's views in the article, and that the Shell group's policy was one of friendly co-operation with Cargill and his friends. Cohen at once proposed "an absolutely frank interchange of views" to heal the differences. Cargill, Hamilton, Finlay and Watson therefore met him and Deterding at the end of November. Cargill insisted on keeping to general points and refused to debate Cohen's accusation against Greenway, of which no more was heard. Cargill had clearly not done badly in standing up to Cohen and his master, Deterding.

Although the merger scheme now faced the hostility of the Admiralty, India Office and government of India, other Whitehall departments still had to make up their minds. In

January 1916 the Foreign Office discussed it at length with Cohen, Samuel also being present. Cohen stressed that a merger would not in any way compromise the status of Anglo–Persian or the latter's product agreements with either the Admiralty or Asiatic. At the same time, he suggested, the Shell group's mature world-wide experience and massive credit standing could be of real value to Anglo–Persian. The group could help to improve the organization of its resources, and to raise further funds without involving the already financially stretched British government. The Netherlands, for instance, was a rich potential source of funds.

The Foreign Office did not wonder if Cohen's honeyed words were proposing a back-door route to securing his objective of control over Anglo–Persian. Instead, it asked how the Royal Dutch interests in Holland might view the consequent loss of control over most of their assets in the group: only the specifically Dutch installations would lie outside the merger. Cohen was at pains to emphasize that Royal Dutch regarded the scheme with a very friendly attitude. Whitehall had only to give its wholehearted approval in principle, he declared, and Deterding would at once travel to the Hague and secure the formal blessing of his Dutch colleagues. Foreign Office opinion remained highly sceptical about these reassurances.

Then the department responsible for both oil and company affairs, the Board of Trade, took up the question. At first it was mainly interested in Cowdray's merger proposals, which it explored with Greenway at the beginning of the year, since Anglo–Persian was a key element in them. The ever sanguine Greenway rather over-played his hand by demanding several measures of official assistance. The Board of Trade was effectively scared off and took up the Cohen scheme instead, which top officials discussed that March with Cargill and Finlay.

Cargill was in one of his blacker moods, emphasizing the disadvantages. Any benefits would be outweighed by the creation of a "huge monopoly with something like government sanction attached to it", and effective control would inevitably remain in Shell hands, however many Burmah Oil directors joined the combine's board. Pretyman, as parliamentary secretary at the Board of Trade, therefore decided to call a conference of representatives from the Admiralty, India Office and his own department, with the experienced Redwood in attendance. They soon agreed to advise ministers that Cohen's proposals should be politely turned down.

That seemed to dispose of the matter for good; then Shell brought out its most powerful piece of artillery, namely Deterding. Having masterminded the merger scheme, he was clearly impatient with Cohen's failure to make headway, and late in June he personally opened informal talks with the Board of Trade. Those turned out to be tough and protracted, but his efforts paid off when in August Lord Harcourt, the acting President of the Board of Trade, produced for a newly formed cabinet oil committee a "Very Secret" state paper on "The Future Control of Oil Supplies".[7]

Harcourt began by stressing how petroleum products had become prime necessities for industry and for land and sea transport. "The problem of supply is therefore no longer merely a commercial question; it is an imperial question of the first magnitude." That was especially so for Britain as the only great power lacking domestic or close-at-hand supplies of oil.

He then considered but rejected the Cowdray plan for an all-British combine, partly because its nucleus would have to be Mexico which was neither politically stable nor a producer of good quality oil. Therefore any effective grouping to secure oil supplies for the British empire must include the Shell group, and Deterding's recent offer to bring its global network of oilfields, tankers and tank installations, except for those that were wholly Dutch-owned, under British control was highly relevant. His proposed combine, with some such title as the Imperial Oil Company, would have both Burmah Oil and the Anglo–Saxon Petroleum Company (a Shell subsidiary) as units. Britain would enjoy a narrow share-holding majority of 51 to 49 per cent and a five to three majority on the eight-man board.

Most agreeably of all for Britain, considerable benefits would accrue with no offsetting disadvantages. Although the scheme could in many respects be construed as involving a restraint of trade, consumers' interests would be fully protected, perhaps by some form of profit restriction similar to the curbs on gas undertakings. Burmah Oil would be safeguarded by guarantees agreed with Delhi against any bid to hinder the development of its oilfields. On the cost side, no outlay of government capital would be required, no introduction of tariffs and no modification of government control over Anglo–Persian: that company would lie entirely outside the scheme, at least for the time being.

Here, then, was a landmark in official thinking. A nation that, theoretically at least, believed in open competition was being offered the establishment of a giant oil trust, to control production of more than 3,000,000 tons a year in the old world, including Burma, as well as in the new; and no fewer than thirty-six refining and distribution subsidiaries. That represented nearly one-twentieth of the year's total world production of about 62,500,000 tons, two-thirds from the United States – yet that new combine's world-wide activities would give it a more than proportionate bargaining strength. Harcourt concluded his paper by seeking cabinet authority to pursue and complete the current negotiations as soon as possible. He must have been pressing for a swift and positive decision so as to confront the still hostile Admiralty with a *fait accompli*.

So believed Greenway, admittedly not the most disinterested party, since he was apprehensive about Anglo–Persian's long-term growth being stunted under the Harcourt scheme. In fact, the Admiralty got wind of the matter just in time for Balfour, the First Lord, to dash down what he called some "stray reflections" before the cabinet oil committee held its inaugural meeting.[8] Above all, Balfour feared the potential power of the new combine to squeeze out smaller rivals and so consolidate its own monopolistic position. At the committee Andrew Bonar Law, Colonial Secretary and one of the few ministers with practical business experience, focused on the logical flaws of the scheme. Either the British members of the proposed board would have watertight powers or they would not. If they did, then Shell would be making considerable sacrifices should the nominees approve policies that harmed its commercial interests. If, on the other hand, the British directors found themselves unable to resist Shell's designs, then the government was risking the loss to Britain of some very valuable assets indeed.

Slade put in his oar with three memoranda which went up to the cabinet as well as to the committee. The haste of their composition barely excuses his slipshod thoughts. To refute the Harcourt paper, he argued that there were no provisions to safeguard oil supplies under wartime conditions (although in the current conflict Shell was providing much of the allies' needs) and also that conflicts of interest could still arise between the British and the Dutch (although the specific intention was to resolve those in favour of the British). Moreover, he felt it would be intolerable to allow a foreign company a 30 per cent stake in British assets, even with apparently watertight control.

For Shell directors (according to Slade) the scheme represented "the successful culmination of a steady policy that they have pursued for the last ten years, by which they have endeavoured, in conjunction with the Standard Oil Company, to obtain virtual control of the oil trade of the whole world." He then put forward what was virtually a rejigged version of the Cowdray plan.

The cabinet oil committee, while coming down on Harcourt's side, did try to meet the Admiralty's more sensible objections by asking if it would accept his scheme subject to certain prior conditions. For instance, the agreements between Anglo–Persian and Asiatic might be renegotiated, and Burmah Oil's shares in Anglo–Persian assigned to trustees so as to ensure the latter's independence. Their Lordships stood firm and Pretyman, with his long experience in oil matters, was given the task of breaking the deadlock. The first thing he did was to sound out Cargill once again.

Cargill, by then in a more robust mood, wholeheartedly commended the Harcourt scheme as being along the right lines and rejected the Cowdray proposals as a sure recipe for disaster. Pretyman then brought together Cohen and Slade for a face-to-face debate. Not unexpectedly, a confused and inconclusive clash of wills followed, and once again it looked as if the scheme was grounded.

Events quickly overtook those parleys. Only a week or so later, in December 1916, Lloyd George came to power as prime minister and very rapidly stamped his own dynamic personality on the entire government machine. In a ministerial reshuffle, Balfour left the Admiralty, and so did his adviser Slade, who became vice-chairman of Anglo–Persian. Pretyman, once again Civil Lord of the Admiralty, was appointed chairman of a new committee on petroleum products. That was to deal with all oil questions, and its problem of coping with the ever worsening shortages gave it no time to consider the merger scheme, even if it had had any useful ideas. Instead, during June 1917, a powerful ministerial committee was appointed specifically for the merger, headed by the new First Lord, the other members being the President of the Board of Trade, the Colonial Secretary and the Secretary of State for India. Only the last-named, Edwin Montagu, was known to be wholeheartedly in favour of Harcourt's scheme; when informing Cargill of the ministerial committee's formation, Greenway added complacently that from what he heard, "the scheme will be buried and cleared out of the way forever."

It was in fact that committee which proved abortive, since it ran counter to the well-established official procedure for dealing with major interdepartmental questions. Normally civil servants clarified the detailed issues involved and then put forward their recommendations to ministers, who would be able to reach agreement in a cabinet committee or in the cabinet itself. To busy ministers, the discordant voices rising from individual departments unnecessarily wasted a lot of energy: the problem of the merger was therefore referred back to Pretyman and his committee. In December 1917 a specialist body to handle day-to-day questions was set up: the Petroleum Executive, under the chairmanship of John Cadman, the mining professor and member of the Admiralty commission who had inspected the Persian fields in 1913.[9]

About that time Cadman chaired an official committee to co-ordinate policy on oil transport in the far east, for presentation to a forthcoming American delegation now that the United States was in the war. Watson was there for Burmah Oil and Deterding and Samuel for the Shell group. Much to the latter pair's disgust, Greenway had sent Duncan Garrow to represent Anglo–Persian. The meeting was held up while repeated requests by telephone were made to Greenway, who declined to attend.

Deterding, having agreed that Burmah Oil should be given priority in the Indian market, then attacked the Anglo–Persian directors for putting themselves on a pedestal, being over-secretive and posing as a semi-government department, thereby creating the impression that they were being unduly favoured by the government itself. The thrust of his attack was undoubtedly that because of its connection with the authorities, Anglo–Persian would do well out of the tanker shortage at Shell's expense. He also condemned the Cowdray proposals for an all-British combine and wondered whether they were meant seriously or were just a bad joke. Watson crept out of the meeting when he had had enough; yet he read the implications of that outburst only too clearly. No merger agreement with the Shell group could hope to be effective unless the question of Anglo–Persian's future role in the oil world were satisfactorily resolved.

Then it was Greenway's turn to proclaim his strongly held views. Perhaps to convince Deterding that the Cowdray proposals were not in jest, in December 1917 he told his shareholders:[10]

> What is wanted is an extension of the policy initiated when the government secured a controlling interest in this company, and the formation of an "all-British company", similarly controlled and free from foreign taint of any kind to deal with the development of oil fields outside of the British isles.

The responsible London press generally approved his ventilating the issue of government control over oil supplies for Britain, and the perils of over-reliance on Persia alone. Yet in other quarters the reaction was less benign.

Samuel in a private letter to Cadman vigorously protested at the epithet "tainted" being levelled at his own company just because it happened to have foreign connections. Typically, he even threatened that if such insults were to continue, Shell might remove its headquarters from Britain altogether. While dismissing that threat as mere bluster, Cadman did admit to civil service colleagues that further official support for an all-British combine would inevitably alienate Shell and Standard Oil, which were after all currently supplying the lion's share of Britain's wartime oil requirements. Cadman therefore advised Walter Long, as Colonial Secretary, and the minister responsible for oil matters, to convene a meeting of departments to hammer out an agreed view on the various merger proposals. He undoubtedly hoped that the Cowdray proposals would be rejected overwhelmingly.

Yet the shock-waves generated by Greenway's statement

brought to the surface all the departmental discords over the whole issue. Questions were asked in parliament, Sir Albert Stanley as President of the Board of Trade replying that the government had a merger scheme "under consideration". Walter Long was furious, the more so when he learnt that the reply had been drafted in his own department, and at once called together an interdepartmental meeting. Alwyn Parker was to represent the Foreign Office, but in his letter of apology made some revealing remarks. Having described the contents of Samuel's letter to Cadman as half true and half false, he added, "I think [that] Mr Greenway, whom I have known intimately for years, is a very unreliable person and that he runs the Anglo–Persian, with the support of Admiral Slade, who follows all Mr Greenway recommends." He therefore suggested that Greenway should be told plainly to submit any future pronouncements of that kind in draft to the government for prior approval.[11]

Walter Long's interdepartmental meeting sought to defuse the issue by getting the government publicly to deny that it was considering the formation of an all-British oil combine and by having Greenway formally reprimanded for having spoken out of turn. Significantly, the meeting did not condemn the Shell proposals, and Stanley at the Board of Trade was convinced that the merger question would spring to life again before long.

Watson was of similar mind. In January 1918 he raised with Cargill the likelihood that once peace returned and the "clamant claims of war" were a thing of the past, the government would "return very quickly to this [the merger] and other ideas for broadening the basis of the empire's future oil supplies": perhaps one of the most pressing of Britain's post-war problems now that oil had become indispensable. So far Watson had been largely on the sidelines over the scheme, but now that he was a director of Anglo–Persian and Cargill's principal confidant he was certain to have a voice in subsequent negotiations.

As Greenway, lately reprimanded by Cadman on behalf of the government, felt himself back on the defensive, he tried to win Cargill's support for his detailed objections to the Shell scheme. Any foreign capital or management likely to be introduced into the proposed combine, he suggested, would in some way contaminate the whole: an echo of the "taint" theme in his earlier speech. It would also lead to 60 per cent of profits on oil produced or consumed in British territory lining the pockets of foreigners.

Cargill asked Watson to comment on these objections. Watson reminded him that the heart of the scheme was British majority control which must be effective enough to keep any foreign interests in check. On Greenway's assertion about profits being lost to Britain, he replied that legislators at home and in British possessions could see that any profits accruing to foreigners would be suitably taxed. Greenway had also suggested to Cargill that Anglo–Persian as well as Burmah Oil would be directly absorbed under the Shell proposals. Watson's impression was that Anglo–Persian would be unaffected as Burmah Oil was no more than a minority shareholder. He also fully backed Cargill's consistent refusal to consider the merger scheme seriously until asked by government. Yet he was more positive than his chairman in stressing the benefits to shareholders and others financially and from what he called "a practical oil strengthening".

In May 1918 Long finally persuaded the war cabinet to set up a body to look into the future of the British empire's oil supplies. The Petroleum Imperial Policy Committee was therefore established, being chaired by the experienced Harcourt and manned by Inchcape, Pretyman, Black and Cadman as well as departmental representatives and some outside business magnates.[12] Among the subsidiary terms of reference were whether the Shell merger scheme should retain its holdings in Anglo–Persian or not. Those holdings had at last been brought specifically into the discussion. Now that they were a key issue, they remained so until it was fought out on the highest level some years later.

Deterding and Samuel, summoned to give evidence before the committee, envisaged that Shell would acquire the whole or part of the government holdings, or be called on to subscribe for additional shares created by Anglo–Persian. They reiterated the arguments they had put before the Royal Commission of 1912 that oil, like everything else, was subject to the laws of supply and demand. To get the oil they wanted, therefore, British and imperial consumers would have to pay the going price. Even so, Shell would grant the whole of the British empire no less favoured treatment than any other market.

Afterwards Cadman sought to assure his fellow committee members that the Shell directors' patriotism and Deterding's openly pro-British sentiments were most valuable assets. He seemed to be gaining much support after the still hostile Greenway and Slade showed that they had nothing particularly new to say.

Slade caused a furore by making some extremely provocative remarks in a memorandum about Royal Dutch's allegedly close relations and trading activities with Germany. Astonishingly, that memorandum, together with a covering note from the First Lord, went up as a state paper to the war cabinet although Slade had been out of the Admiralty for months. The highly embarrassed Sea Lords had to put up a supplementary memorandum lamely explaining that they had intended merely to endorse Slade's not very original remarks about the Persian oilfields and prospective Mesopotamian concessions being of crucial importance to Britain.[13] Between them Greenway and Slade had finally dished any plan for a merger that would exclude the Shell group.

Then it was Cargill's turn to attend the Harcourt committee. He had already drafted some disjointed notes of his own, which Watson with his usual skill transformed into a positive and wide-ranging document. It stressed that Britain's very first priority must be to carry out a systematic geological survey of the whole British empire and to see that all discoveries were then tested by the drill. In India (including Burma) it was vital to sweep away obstructions to efficient oil raising such as the reservation of blocks against Burmah Oil. As to the Shell merger scheme, it merited government's most earnest consideration since, contrary to some people's views, British financial control over the proposed combine could increase both oil resources and employment and production generally within the empire.

Having presented that document to the committee, Cargill was then questioned on it. The prominent industrialist Sir Harry McGowan, who saw in him "a man with imperial ideas", elicited the reply that the empire would indeed benefit from really effective British control of the combine, something well worth while even if Burmah Oil's holding in Anglo–Persian had to be sacrificed to achieve it. To help clinch that control after the merger, he suggested that Shell should appoint an additional managing director, working full-time. He even put forward a name that was too secret to be committed to paper; that turned out to be none other than Watson. Inchcape, already well acquainted with Watson's strong personality on the Anglo–Persian board, judged that there was no likelihood of Deterding even considering him for such a position, and that quite impracticable notion quietly faded away.

Then in October Deterding came back to the committee with some rather tough conditions for a merger, at last bringing Anglo–Persian specifically into the scheme. Shell would buy half the government's holding of ordinary shares in that company at par, although they would have fetched in the market three or four times their nominal value. The other half should be sold to Burmah Oil at the current market value. The government would be allowed a royalty on Anglo–Persian's output. In return, Royal Dutch–Shell was prepared to become entirely British in terms of voting power, except for its activities in Dutch possessions. He later raised his terms by proposing that Shell might buy enough shares, at a price to be agreed, to give them a 49 rather than 33 per cent stake in Anglo–Persian. He disclaimed any wish to interfere in its management and asked for no more than an advisory seat on the board.

These audacious proposals cost Deterding the goodwill of Cadman, who formally requested the committee to reject the proposals. However, as soon as the armistice of November 1918 was signed, Deterding relaxed his tough bargaining stance. Harcourt and his committee members inferred from this tactical move that "doubtless the Netherlands would prefer the friendship to the hostility of the greatest naval power in the world": a delightfully nineteenth-century sentiment that might have graced the despatches of Canning or Palmerston. In fact Deterding was after much bigger game. As mentioned in Chapter XIII, the 1914 Anglo–German agreement on Mesopotamia had given the D'Arcy group a 50 per cent share in the concession, to Shell's 25 per cent. After the military defeat of both Germany and Turkey, Deterding was resolved to win for Shell parity with Anglo–Persian, as well as the exclusive right to manage the Mesopotamian operations. The Harcourt committee now offered to meet both his demands, on condition that he brought all the Shell interests, except purely Dutch ones, under British control. That would be done by rearranging shareholdings and therefore any merger with Burmah Oil and Anglo–Persian would no longer be necessary.

That accord was initiated by Harcourt and Deterding in March 1919.[14] Whitehall was on the brink of an astonishing victory at no real cost to itself. Shell and its world-wide assets would be once again truly British, while Mesopotamia was as good as won because the Turkish Petroleum Company would operate there as a British company. Since neither Burmah Oil nor Anglo–Persian would be linked with Shell in any way except through their agreements, Admiralty and India Office objections fell to the ground. As to Deterding, here was his finest hour. Not only had he secured a large interest in Mesopotamia at Anglo–Persian's expense but more importantly, as a recently naturalized subject and therefore more British than the British, he, and through him the Shell group, had now attained a standing with the government that they had never enjoyed before.

Yet if the principals were more than content with the Harcourt–Deterding accord, Greenway and Watson were at one in condemning them. Greenway, particularly riled at having forfeited to Shell the opportunity to work the future Mesopotamian deposits, protested to Cadman as head of the Petroleum Executive, asking why the Mesopotamian and Persian operations could not have been carried out by Anglo–Persian under unified control in the interests of economy and efficiency: after all, he argued, his company had mature technical knowledge of the whole region. To his fellow-directors in Anglo–Persian, Greenway was more forthright: "[It] looks very much as if the nation is selling its birthright for a mess of pottage."

Watson privately expressed his views to Finlay Fleming & Co. in Rangoon. A stiff price, he stated, was being demanded of Britain to secure a "paper" control of the Shell group, and he devoutly hoped that what he felt to be a gigantic bluff would fail. As in the succeeding months he saw the Harcourt–Deterding accord founder, ostensibly over very vocal Dutch objections to various provisions, he was more than ever convinced that the only sure solution would be an outright merger between Burmah Oil and Shell. It was the quest for that solution that was to dominate his working life during much of the next five or six years.

Notes

1 The surviving company papers on the merger scheme have been collected into two files, with correspondence and memoranda running from Cargill's letter to Slade, 30 June 1915 to Cohen's letter to Watson, 23 December 1924. See also Cargill's comments in 1924 AGM, *Glasgow Herald*, 11 June 1924.

2 PRO ADM 1/8446/13: memo by Sir F. W. Black (Director of Navy Contracts), 20 July 1915. This file runs to 23 August 1915 (with two memos of February 1916) and partly overlaps with FO 371/2426 for 1915 and 371/2721 and 382/1096 for first half of 1916. ADM 1/8537/240 basically runs from July 1916 to September 1918.

3 Ibid. PRO ADM 1/8446/13: joint opinion of W. H. Upjohn K. C. and H. O'Hagan.

4 J. A. Spender, *Weetman Pearson, First Viscount Cowdray 1856–1927* (1930), pp. 202–3. See also Dutch edition of Gerretson, *History of Royal Dutch*, V, pp. 104–5. Black's remarks are in the memo of 20 July 1915, see note 2 above.

5 See (unpublished but printed) "History of Ministry of Munitions" (1922), in Bodleian Library, Oxford, VII, p. 136, and Henriques, *Robert Waley Cohen*, pp. 201–7.

6 In *Syren and Shipping*, 27 October 1915. See Henriques, *Marcus Samuel*, pp. 611–12.

7 The Board of Trade (Harcourt) memorandum can be most conveniently seen (in the context of earlier and later enquiries from the Admiralty Committee of 1911–12 to the 1916–17 committees) in PRO FO 368/2255, III, Introductory pp. 5–14 and Appendix A (see note 11 below). Regrettably, the Board of Trade archives on the question (and on many interesting questions in the nineteenth and early twentieth centuries) have disappeared. Deterding's "tentative and informal" discussions with the Board of Trade are mentioned in FO 371/2721, Foreign Office to India Office *et al.*, 27 June 1916, and with the Admiralty in ADM 1/8537/240, memo of his and Cohen's meeting with Director of Intelligence Department, 13 July 1916.

8 For Balfour's "reflections" see PRO CAB 37/154, 3, 18 August 1916, and for Slade's three memos see ibid., 16, 6 September 1916.

9 The complex history of government bodies, including the Petroleum Executive (December 1917), to deal with oil affairs is usefully unravelled by M. Kent, *Oil and Empire*, Appendix VI, "Note on Government Mineral Oil Organization during the First World War", pp. 185–8.

10 APOC AGM, *The Times*, 4 December 1917.

11 A. Parker to J. Cadman, 21 December 1917, PRO POWE 33/42.

12 PRO FO 368/2255: H.M. Petroleum Executive, "Negotiations Regarding the Petroleum Policy of H.M. Government I Report and Proceedings of the Petroleum Imperial Policy Committee, 29 May 1918–10 February 1919." See also PRO POWE 33/64, 65, 68, 69 and 76 January–May 1919.

13 PRO FO 368/2255, Appendix D: memo by Slade, 30 July 1918. For the extraordinary rumpus arising out of this memo and Slade's intemperate accusations that the Petroleum Executive was "very leaky" see PRO POWE 33/45 7 September–7 October 1918. cf. disclaimer by Sir E.

Geddes (First Lord), 17 September 1918, CAB 21/119. Watson had a field day scribbling his points of disagreement over the company's copy of Slade's memo, and maintaining that British financial control of foreign oil assets plus physical control of the sea lanes *would* effectively safeguard Britain's oil interests.

14 The Harcourt–Deterding agreement of 6 March 1919 is in D. J. Payton Smith, *Oil: A Study of War-Time Policy and Administration* (History of Second World War, UK Civil Series, 1971), pp. 28–31 and in Kent, *Oil and Empire*, pp. 178–82.

CHAPTER XVII

Into the Post-war Era 1919–24: (1) The Watson Touch

On 2 January 1919 R. I. Watson sent John Cargill a memorandum setting out the company's current position and its prospects for the future. By a mischance no copy of the memorandum has survived. However, as a number of points were discussed in correspondence, its burden seems clear.

In the decade between Kirkman Finlay's death in 1903 and the outbreak of war, the company had made very good progress, with refinery throughput more than trebled to just over 4,500,000 barrels and profits from Rangoon almost trebled to £971,000. The Persian oilfields had been successfully opened up and powerful world-wide rivals fairly well contained.

The main snag was that wild-cat drilling in new areas in India and Burma had been disappointing so that, at existing depletion rates, it was probable that the company would have virtually exhausted all its fields within the next twenty years or so. Moreover, production costs would soar as wells went deeper, so that there must be the most rigorous economies, particularly on capital projects. Indiscriminate activities in the past had led to much waste, and everything from now on must be carefully and deliberately planned.

Refining methods above all required attention, both to conserve the increasingly scarce crude and to obtain the highest possible returns on those products in greatest demand. As the existing plant was obsolete, two basic decisions would have to be made: to consider concentrating all the refineries on one site, and to choose between the dephlegmation and atmospheric processes, arguments over which had aroused much rancour in recent years. While considerable sums would have to be expended, the prospects for economies were excellent.

As to marketing, the current kerosene agreement with Asiatic badly needed revision to bring it more in line with post-war conditions. There were also the consequences of Anglo–Persian's arrival on the world oil stage: in the current year of 1918/19 the amount of crude produced in Persia reached 1,000,000 tons for the first time and hence already exceeded total production in Burma.[1] Those two issues, and others affecting Burmah Oil's relations with Asiatic, Anglo–Persian and others, will be considered in Chapter XVIII.

Many defects in organization impeded efficiency. In particular, the "triangle" of communications between Glasgow, London and Rangoon (not forgetting Calcutta) multiplied work unnecessarily and created opportunities for confusion. In the years that followed, much of Watson's efforts were devoted to simplifying the lines of communication in one form or another. Yet, for all his far-reaching plans, in January 1919 he was still the most junior director. How far would he be allowed to put his plans into effect?

Here he was helped by the virtually clean sweep of top management during the war. The last one to depart was Hamilton; when his health broke down early in 1919, it was decided to transfer the stores and plant department together with the engineering department to London. Most agreeably, Watson was anxious to have these departments immediately under his control while Cargill was glad to let them go.

Hamilton's departure also left the way open to the appointment of a managing director. That was not to be Sir C. K. Finlay, who was still serving at the Ministry of Food and had little contact with company affairs. He could of course have resigned from the civil service but he did not want to return full-time to the London office. His heart complaint, although no drawback in undemanding work, made it risky for him to undertake a job that had driven his father to suicide and which had become even more onerous since 1903. In any case Watson, as the ideal candidate, was already running things in London and was perfectly capable of shouldering all the burdens involved. Finlay therefore remained a non-executive director until his death in 1937.[2] Cargill often consulted him on matters of policy, but at this point he passes from the mainstream of the company's affairs.

Watson was appointed managing director as from 1 January 1920; he was just 41. By coincidence, on the same day Cargill was made a baronet in the New Year honours, for public services. Just at the time when he ceased to be the most influential member of his organization, Cargill was confirmed as a man of consequence in the eyes of the outside world. As well as serving on the boards of various tin, rubber and investment companies, he was very active in the civic affairs of Glasgow, including its Chamber of Commerce and the university. Those activities brought him much public esteem but so ate into his time that he was henceforth really a part-time chairman, concerning himself either with trivia or with matters of the highest policy.

The tone of his letters contrasts sharply with those written by Watson. Cargill was typically benign, sentimental and at times emotional: "very disgusting" or "really heartbreaking", he would comment on some instance reported home of a rival's malpractices or a piece of sheer bad luck. It was not for him as chairman to issue lengthy and detailed directions to subordinates. Instead, he did his best to keep a paternal eye on those in the east, particularly over personal and welfare matters. Once a month he travelled down to the London office, a chore to which he never looked forward. Every three or four years – except during the war – he made the trip to the east, where his tasks were to an increasing extent ceremonial.

Watson, on the other hand, was a man of mind rather than heart. Lacking Wallace's hard precision of thought, he managed to see all the varied sides of issues under his consideration. Whether dictating or scrawling comment in his execrable handwriting across documents that came to him, he seldom left a stone unturned or avenue unexplored, and it is not surprising that those at the other end often failed to grasp his meaning. Although mellower than he had been, he was at times abrasive in correspondence with Finlay Fleming & Co. He respected the abilities of the general manager, Adam Ritchie, but never shrank from pointing out deficiencies in the agency's running.

Above all, Watson sought to enforce a proper chain of command. Little by little he weaned Ritchie from corresponding with Cargill direct. In 1921 he tactfully took him to task for sending the chairman a report about a recent visit to the fields, and laid down that all writings must be sent to the London office, so that the relevant departments there could read and act on them. "This, of course," Watson added, "does not mean that when you have some reason to trouble Sir John personally you should not write to him privately on unusually important and confidential questions": an eventuality clearly expected to arise very seldom.

The following year Ritchie discontinued his weekly private letter to Cargill, who admitted to his annual meeting of 1923 that thanks to Watson's able management his position as chairman was one "nowadays of comparative ease". No friction ever arose between the two top men in the company over these changes. It never crossed Watson's mind that he might be empire-building. He merely detested inefficiency of any kind.

The consequences to Burmah Oil of Watson's promotion to the managing directorship were momentous. He firmly re-established the day-to-day control of the company that had been basically lacking ever since the first managing director, Kirkman Finlay, had died in 1903. He also had the aptitude to make good use of that control. He possessed formidable brain power, memory and application to work, was familiar with the east, and had had his horizons greatly broadened by serving on the wartime Petroleum Pool Board and by exposure to the large ideas of Robert Waley Cohen.

Once the stores and engineering functions had been transferred to London, he had the technical people he needed on the spot, particularly Andrew Campbell the advisory chemist and John Gillespie the engineer. Sir Boverton Redwood, former consultant to the company, had died in June 1919, aged 73, having been overworked on government business throughout the war. He had rendered impressive service to the infant company as a nanny, but it was now of age and no successor was appointed.

Watson's prime need in the London office was a deputy. Having for once consulted the board in advance, he asked Ritchie and his second-in-command, G. C. Whigham, to decide between them who should come home. Unlike the energetic and sporting-minded Ritchies, Whigham and his socialite wife had never really taken to Rangoon. He therefore returned to join the London office and the local London board at the end of

1919. After Hamilton's death he was elected to the Glasgow board, and became assistant managing director in 1923.

The registered office had to remain in Scotland, as did the share registers and the accounts department, the last-named not moving down until 1941. After the retirements of Adamson and Hamilton, the Glasgow office seems to have gone to seed; but Watson was in no position to wage war against the slackness there as he did against the Rangoon agency. Except on full-dress occasions such as the eve of the annual meeting, Glasgow board meetings consisted only of Cargill, his brother David – who was concerned mainly with non-oil matters including those of Cargills Ltd – and one or two non-executive directors. The board was informed of some, but not all, of Watson's decisions as they were made, and it transacted routine business such as authorizing donations. As an ordinary director Watson had never bothered to attend, regarding any trip north as merely a "jaunt" that would take him away from more pressing work.

Watson's thinking dominated the company's affairs from 1920 on. His watchwords were economy and efficiency, and he expected full co-operation from Rangoon, where all the company's trading profits and five-sixths of its total earnings came from. That March he instructed Ritchie to send home regular analyses of income and expenditure, to show total and average costs per gallon on the different operations such as products, the pipeline, refining, shipping and marketing. Each operation had to be broken down into labour, plant and stores, and interest on capital. In the fields he set up an engineering department to cost all projects there, but taking account of the economic and other advantages of such projects.

By August he was scrutinizing all items of capital expenditure in Burma, and their cost as well as superfluousness appalled him, bearing in mind the ephemeral nature of oil production. Although the Yenangyaung field was well past its prime, he demanded, why were the old functional buildings there being scrapped wholesale and replaced by expensive permanent structures? And the same as to men. Why did a draughtsman have to be sent from Britain at great expense when a local technician could have been trained up?

Rangoon replied that many buildings were dilapidated or worse, as no major repair work had been possible during the war. Watson, hot in his resolve to cut really deep, was merely strengthened in his conviction that Finlay Fleming & Co. was not pulling its weight as it should do. The agency's income was mainly from its commission on Burmah Oil sales, and he must have known how much of Cargill's time was taken up with post-mortems into the losses made by most non-oil departments in Rangoon.

Among Watson's main *bêtes noires* were the technicians, both in the refineries and the fields. "There is no greater spendthrift than the purely technical man," he declared with the sure authority of hard-won experience, "and he must be controlled." Yet his enthusiasm for pursuing economy to the extreme was bound eventually to come a cropper. Towards the end of 1921 he began to ask questions about the perks enjoyed at the company's expense by employees out east. He demanded details of free or preferentially priced supplies of company products and of servants provided without charge to employees. He was startled to learn that assistants in each Rangoon chummery (or bachelor quarters) were collectively allowed the services of a butler, two sweepers and two gardeners as well as free electric light. Such perquisites were unheard of in his days out east, he complained to Ritchie, and all should be abolished.

Rangoon insisted that they were fully justified, the butler because someone had to be responsible for the firm's property, including the furniture and the crockery; the others for comparable reasons. As to electric light, it emerged that Watson himself had authorized its provision in lieu of earlier free supplies of kerosene. Watson backed down with the remark that he just wanted to ascertain the exact situation, and then ban any extensions of the "free list" without his special permission. He managed to get in a dig later about employees in Burma being "petted and pampered".

To aid his economy drive, for the first time he brought in a management consultant. To be sure, W. L. Mackenzie had in 1915 looked at the fields set-up, but very much from the viewpoint of a professional engineer. Now Watson chose Albert Cathles from the accountants Price Waterhouse & Co. Cathles was the author of *The Principles of Costing* and already enjoyed a wide reputation as a costing and systems expert.[3] A taciturn and very erudite Scot, he had been awarded the OBE for his wartime work of setting up logical systems of organization and control in munitions factories.

Watson asked him to study both the refinery and fields organizations and the company's costings systems generally. The report he submitted late in 1921 – now unfortunately lost – concluded that the costing procedures were satisfactory, but that functions in the refineries and the fields needed to be decentralized. The technical men, such as works or fields managers, could then become free of accounting, storekeeping and administrative responsibilities. The refineries, for example, needed a trained accountant to deal with costs and other financial matters. Watson at once made a start with implementing these recommendations.

Well-conceived or not, Watson's economy drives were none too premature. In 1913 gross profits per barrel of refinery throughput had been 21p, not so different from results in the early years. By 1918 they had risen to 61p and in 1919–20 to an average of 89p, when costs generally were about double the 1914 level. Then, with the onset of the depression, in 1921–2 profits fell to 51p on average and in 1923–4 to less than 50p. Meanwhile, expenses had risen alarmingly, drilling costs in 1920 being 130 per cent up on 1918. Admittedly, the very depleted stocks of new materials such as piping and drilling equipment had to be built up at a time when their prices and transport costs were unprecedentedly high. Despite much co-operation from staff, Watson knew there could be no appreciable reductions in costs for some time to come.

Another bar to economies was the poor quality of the drillers. Many went home as soon as the war was over, and their replacements were sometimes transparently seeking to escape from the prohibition laws just enacted in the United States. Even more difficult was finding experienced Americans who could be groomed for senior posts such as fields superintendents. Watson candidly admitted that those with genuine technical and administrative skills would scarcely wish to go abroad when they could find well-paid jobs in relative comfort at home. As senior staff in Burma had always been promoted from inside, those who had spent years solely in the company's service had few up-to-date economy-promoting ideas to contribute. In 1921, therefore, he decided to release senior men for regular visits to oilfields in the United States and Europe to learn about the newest technology.

Yet he knew that the only long-term remedy was to train British and Burmese drillers. By the end of 1922 he had set up a school of drilling in the fields for a preliminary batch of eight Scottish ex-Royal Engineers. They proved some of the best recruits the company had ever had, and some in due course took over from the Americans as fields superintendents. Within a year or two they successfully introduced rotary drilling, which had never before been made to work in the province. That recruiting scheme led to Burmese trainee drillers being taken on as well.

The heavy fields expenditure of more than £1,000,000 a year was on a scale to diminish profits from Rangoon by nearly a third. To see what further economies might be possible, in mid-1921 Ritchie investigated the costs of each field in turn. The bleak conclusion was that few extra reductions could be wrung out of them. Watson knew that the kind of radical reorganization he hankered after would have to wait until Seiple's departure, for Seiple, it will be remembered, had prevented the fields agents from having an effective say in the running of the oilfields.

When Seiple did retire at the end of 1923, Watson avoided any repetition by not appointing a successor but instead strengthening the fields agent's authority. The superintendent in each field would be responsible to the sub-agent there, who would take his orders from the agent. That reform was still being carried through when Watson made his trip to Burma in 1924–5. It was his first visit to the fields since 1905, when Jacobs had ruled supreme and virtually unmindful of the company's wishes. Yet even these changes in organization were dwarfed by his major decision to transfer to the London office direct control of fields operations.

He also transferred to London marketing policy in India and control of various other functions in the east. Keeping constantly in mind his threefold objectives of speed, efficiency and economy, he and his office would be able more effectively to devise broad policies and act as "helpful and friendly, if sometimes pungent, advisers and critics". That rationalization he considered to be "among the best fruits" of his visit, and, he added, "even were it the only one, then time occupied would have been well spent."

A less tractable problem was that of the geologists, for no amount of reorganization could drive them into finding oil if none were there. The old-established techniques of studying geological formations were still in use, for the era of geophysical surveys lay in the future. It irked him that his geologists were as "frail as others elsewhere of the same persuasion". Sometimes they appeared considerably frailer, as when in 1921 Indo–Burma struck oil in Lanywa, north-west of the Singu field where it was intersected by the Irrawaddy. Dewhurst as chief geologist and his staff had already declared confidently that oil would never be found there in paying quantities, and Ritchie was as shocked by the news as they were. Watson's response was unusually low-key, but in contrast with Cargill he declined to waste energy on worrying about matters he could do nothing to remedy.

There was a great deal he could do about refining, in particular by pushing through the Refinery Concentration and Reconstruction Scheme. The controversy as between Allan's atmospherics and Campbell's dephlegmators had been temporarily halted by the war. Yet even when peace returned, Watson and his technical staff in London did not hasten to make up their minds. Only lengthy tests of each process under working conditions could objectively prove which was the better and – as important in this matter – convince both the strong-minded and abrasive Campbell and the no less determined Allan.

Having ordered a dephlegmator unit, Watson arranged for an expert to test its performance alongside the atmospheric bench in Rangoon. That expert, Anglo–Persian's distinguished chemist, Dr Samuel Auld, very soon convinced himself that the atmospheric method was well ahead of its rival in both economy and efficiency. Allan, having recently been appointed works manager, was forward-looking, scientifically well qualified, and concerned for good administration. Whereas his predecessors had insisted on signing almost every document leaving the works, he delegated to his subordinates and thus gave himself time for long-term planning and basic research. Together with his chief engineer, James Moore, he devised a series of patented improvements in wax manufacture, such as a cooler and stove; these the company later adopted.

Early in 1921 Watson, with Cargill's approval, made his crucial decision: to adopt the atmospheric method. Three new benches would be installed, all at Syriam, so that Dunneedaw could be turned into a store for refined products. Allan reckoned that the total cost of £700,000–£1,000,000 should be recouped in savings and increased revenue within eighteen months, partly by cutting refinery labour by more than a half.

Although the entire scheme was a monumental undertaking, it could not move fast enough for Watson. By the end of 1922 virtually no plant had been ordered, and the London office was still largely in the dark. "Please have no anxiety as to our ability to tackle and complete the work here," Watson told Rangoon ominously, and he instructed Moore to travel home at once, "fully armed with all relative data including plans, specifications and estimates to enable the scheme to be completed here without further reference to your side".

Here was centralization with a vengeance. No sooner had Moore arrived than he was in regular conference with Watson and Gillespie, to arrange the ordering of the plant required. Then Allan came home with one of his new inventions and, as Ritchie was on home leave as well, Watson could mastermind the whole reconstruction scheme without ruffling anyone's feelings. Then in 1924 he saw on the spot how it was coming along, and commended the efforts being made to keep throughput at normal levels during the operations. Although completion was still well in the future, he looked forward – much as Kirkman Finlay had done in 1902 – to "a refinery second to none in excellence of layout, technical modernity and compactness". Not until the 1929 general meeting was Cargill able to report that the scheme had been satisfactorily completed and was yielding the planned benefits and economies.

Watson also brought electric power to the fields, the scheme being completed in 1922.[4] As with the refinery reconstruction, there were knotty technical problems over which he readily confessed his ignorance, but by then he was skilled in making the experts dance to his tune. It was in 1918 that he put the electrification scheme out to tender, with the help of John Gillespie and of Anglo–Persian's Christopher Dalley who had done a similar job in the Persian fields. He accepted the bid of British Thomson-Houston in preference to its rival, the Westinghouse Electric Company Ltd. Then he engaged McCarthy-Jones, who had just resigned from the former, to oversee the scheme from Britain, and W. S. Toplis from the latter to reside in Burma and supervise the plant assembly on the spot.

Watson informed the board – without asking for its prior agreement – that the cost would be about £1,000,000 and reckoned that it could be completed within three years. Cargill took an appropriately sombre view of the likely cost and running expenses, which would not be known certainly until after completion. "Estimates are just about as dangerous and uncertain as 'gushers'!" – flowing wells that might be prolific for years or die off in a few days. Watson refused to be rattled. He knew it was right to go ahead, particularly as the expected physical saving of oil would help prolong the life of the fields; yet he had to make formidable efforts to keep costs in check.

Inevitably, the uncertainties of the post-war world led to delays and snags. Various strikes broke out among home suppliers of equipment: then as the buildings started to go up, local labour and materials were found to be unreliable. Technical problems arose as well. The essential water cooling system had to be carefully planned as the level of the Irrawaddy varied so much depending on the season. Above all, Watson took a calculated risk in choosing as advisers two men from rival companies and with differing temperaments. "The object," he later explained, "was to get the best of the knowledge and the brains and the experience of both sides and to make use of the natural commercial rivalry of the two concerns in putting these men as much as possible against each other in an associated capacity."

In spite of everything, they succeeded in completing the scheme ahead of time, and the official opening took place in July 1922. A steam turbine plant, rather than the technically more efficient diesel equipment, had to be used as it could burn natural gas which would otherwise have gone to waste. At last the heavy consumption of crude at the fields, by then as high as 22–26 per cent, was almost entirely eliminated. The total cost worked out at £835,000, written off against profit at the rate of £100,000 a year.

The scheme was memorable by any standards. For the first

time anywhere in the world, electric power was being used for all operations in an oilfield. Again, for the first time in the company's history a supplying firm had been fully consulted in the planning of a project. Earlier suppliers had been expected to provide what the company chose to order, regardless of whether it was suitable or up-to-date. One bonus, which the observant Watson noticed on his visit in 1924, was how quiet operations had become, compared with the constant racket of the former steam engines. Pumping wells, inactive for most of the day, now needed power only when they had to be pumped.

To be sure, minor problems arose from time to time. When Watson once wrote to the head of the electrification department at the fields, demanding an explanation for the too frequent power cuts on one of the lines, he was told that snakes were in the habit of shinning up the pylons, mistaking them for palm trees and the insulators at the top for the bats on which they preyed, thus electrocuting themselves. Watson was irate at what he thought was a very lame excuse, but agreed that metal discs, similar to those put on ships' cables to deter rats, should be fastened to the pylons' legs.

In January 1921 the company acquired the Assam Oil Company.[5] Early in 1914, H. S. Ashton as the local London director responsible for liaison had proposed merger terms to his Assam Oil colleagues, but subsequent talks had been impeded by the outbreak of war. Then in 1919 serious negotiations were precipitated by something quite unexpected: an official refusal of Burmah Oil's application for concessions in Upper Assam.

Ever since the limited life of the Burmese fields had become apparent the company had turned its attention to mainland India, including Assam (see Chapter XIV). When its application there was turned down, Burmah protested strongly to the government of India on the grounds that other companies were holding on to concessions in Assam without troubling to work them. That was obviously aimed at Assam Oil, which soon afterwards received a sharp reprimand from officials in Delhi. Never at the best of times a very go-ahead company, it was making so little profit that it could not have afforded to exploit those concessions; acquisition by Burmah Oil could therefore be the answer to its problems. After much bargaining a deal was concluded, Burmah Oil in the end securing 90 per cent of the share capital. It was during these negotiations that Watson came to know Albert Cathles, who subsequently acted as management consultant in Burma. From the previous board, only Ashton was retained. Cargill became chairman, and among the other directors were Watson and Whigham.

Watson at once threw himself heart and soul into making Assam Oil a genuinely profitable and high-yielding producer. He had a completely free rein: Cargill did not interfere in any way and the Burmah Oil board members automatically ratified every step he thought fit to report to them. Within weeks of the take-over, he instituted letter books for Assam Oil affairs, organized on the same lines as those introduced for Rangoon in 1912. From Finlay Fleming & Co. he sent over an office man, an accountant and a geologist to the Assam Oil site at Digboi. This team discovered that the working conditions and equipment were extremely primitive. Very soon the first costing system was introduced with the object of pricing all products remuneratively.

Watson soon drew up his policy directives and then left it to Ritchie, in local control of those seconded from Rangoon, to lay down the detailed plans. The next priority was to overhaul prospecting and drilling operations. Dewhurst as chief geologist carried out some systematic tests, the prelude to an intensive drilling programme in 1922. Once much larger reserves of crude oil had been proved, some heavy expenditure on refining and marketing could be justified. Until then, he had to be content with emergency improvements in the haphazardly laid-out refinery; those did yield some very useful increases in capacity and also cost savings.

Yet he recognized that the major reconstruction needed at Digboi would take some years and consume a great deal of money. He was prepared to make massive loans from Burmah Oil funds for as long as they were needed: by 1924 they were to total £320,000 and four years later no less than £1,000,000. He drastically wrote off fields expenditure against revenue, so that declared profits of Assam Oil in the 1920s were too low for ordinary or even preference dividends to be paid. He took the general manager and his assistant off all routine duties to force them to concentrate on policy matters, including expansion and improvement.

During his tour of 1924 he inspected Digboi with much interest: there he could see both the relics of the pre-1921 chaos and the remarkable progress made despite the climate that was disagreeable for much of the year. As generations of Burmah Oil employees who subsequently served there will remember only too well, it was stiflingly hot during the summer, August being the most unpleasant month although, like Rangoon, tolerable in the cold weather. One victim probably exaggerated when he feelingly described the weather as "always abnormal".

Watson's reward came in the early 1930s when production at Digboi began to soar and contribute markedly to India's requirements of petrol and kerosene: it proved of inestimable value when the fields in Burma were destroyed before the Japanese invasion of 1942. Here was the essence of his reforms at their most far-reaching: totally logical, carefully thought out, all-pervasive and looking above all to the long term.

Among matters less directly to do with oil, the establishment of a subsidiary to manufacture tinplate for kerosene tins very much occupied Watson's mind once he had signed the preliminary accord with Tata early in 1918. A new company, the Tinplate Company of India, was established in Calcutta as a joint venture with capital equivalent to more than £700,000 of which Burmah Oil contributed two-thirds and Tata one-third. The company's general manager in India became chairman and Shaw Wallace & Co. the managing agents.

The planning of the factory, near the Tata steel works, is interesting because it illustrates the traditional Watson touch applied to rather different circumstances. He required the factory buildings to be of a reasonable standard but without "frills", and he insisted on the contractors for the plant, some British and some American, tendering on an equal basis, to allow him to cross-check one with the other. His original intention had been to copy a Welsh tinplate works exactly. Then he heard of a novel type of rolling plant in the American state of Indiana that would meet two problems common with India: a shortage of skilled men, and high temperatures in the summer. The United States technology was more capital-intensive than in Wales, but labour productivity was over three times as high as in Britain. Watson accepted that, but looked to Wales for expertise in the tin-finishing operation. By that combination, he declared, "I feel confident that we are getting the best of the old and the new world."

Much to the dismay of all concerned, costs began to get out of hand, and, to crown it all, Tata refused to put up any more money. Cargill could be relied on to contribute his mite of gloom during a bad spell: from the east in 1921 he wrote dejectedly to Watson, "Things seem to be getting into an awful mess in connection with the [tinplate] company." Even more depressing was that the normally strong-nerved Ritchie had concluded that the tin works could never be made to pay and was urging the directors at home to "make up their mind to cut their loss". In March 1921 Watson, in Cargill's absence out east, therefore called a full board meeting in London, bringing south the Glasgow directors at very short notice. To put pressure on suppliers to pass on reductions in the prices of materials, then just beginning with the onset of the slump, he obtained the board's authority to withhold further capital payments from the tinplate company unless it kept within the estimates. It duly came into line, and by 1924 Burmah Oil had sunk about £1,350,000 in it.

The first tinplate was rolled in December 1922. A year later all six mills were in operation. Cargill, his worries dispelled, was then able to report to his shareholders, "A new industry has thus been successfully established in India with our assistance." When they visited India in 1924 he and Watson inspected the works, but it was Watson who went into every aspect of the operations with his usual thoroughness. On the whole he was gratified with the new company's success. By then it was meeting nearly 40 per cent of India's total tinplate requirements: two years later its output for the first time exceeded imports.

In some ways Watson resembled the first chairman, David Sime Cargill, combining breadth of vision with an unquenchable passion for detail. That combination was undoubtedly seen at its most formidable during his first five years as managing director. In his vigorous early forties he had the authority and the energy to carry out the series of internal reforms so vital if the company were to survive in the unfriendly post-war world. At the same time he was becoming an increasingly influential figure in world oil affairs, as Chapter XVIII will explain.

Notes

1 See useful table of Anglo–Persian's profits, production, etc. in B. Shwad-ran, *The Middle East, Oil and the Great Powers* (2nd ed. 1959), pp. 162–3.
2 AGM 1937, *Glasgow Herald*, 5 June 1937.
3 A. Cathles, *The Principles of Costing* (2nd ed. 1924).
4 C. H. McCarthy-Jones, "Electricity Applied to the Winning of Crude Petroleum, with special reference to the Yenangyaung Field, Burma", *Institution of Petroleum Technologists*, 10, 1924, pp. 115–55, especially pp. 139–48.
5 See ch. IV, note 3.

CHAPTER XVIII

Into the Post-war Era 1919–24: (2) Competition and Collaboration

In his external relations, R. I. Watson's overriding concern was to build up links with other oil companies of kinds that were appropriate to the conditions of the post-war world. As he had little or no leverage over such outside competitors, even his pertinacity brought only variable returns. To set off against his successes, most notably a very close collaboration with the Shell group by the end of the period, some rivals such as Standard Oil, and nearer home Indo–Burma, were as much at arm's length as ever. Indeed, even with Anglo–Persian many problems were left unresolved until a more general settlement took place in 1928.

After the wartime ups and downs with Asiatic, Watson believed that the existing arrangements over kerosene were unsatisfactory in two ways. First, Burmah Oil was still barred from producing as much kerosene as it wished, although it could expect substantial increases in output once the refineries were reconstructed. Second, whilst it was selling its full quota at the maximum price, many dealers were exploiting the shortages of the day, and contravening government decrees by overcharging.

At the end of November 1918 he issued an ultimatum to Asiatic threatening to suspend all kerosene deliveries after May 1919 unless Cohen signed a new agreement that would meet two conditions: Burmah Oil's output must no longer be limited, and its permitted sales within India must be at least equivalent to those of 1914 plus an increase based on the growth of sales between 1906 and 1914.

His second condition was that consumers must actually pay the prices laid down by producers. To achieve that, he had his own bold but simple device: the two companies would pool all their kerosene supplies for India. Those would then be sold at a weighted average of actual cost at Indian ports, the weight being determined by their relative contributions to the pool. Burmah Oil's production would go into the pool, the maximum price of 1905 being maintained, and Asiatic would have to make up the shortfall from Borneo at the world price, then two and a half times as high.

The attitudes of officials in Delhi to the proposed pool were of course crucial. Were they to reject it, the wartime problem of insufficient supplies for India's needs would persist. However, as in 1905 and later, the promise to maintain a maximum price was an invaluable bargaining counter. Delhi did concede the company's right at any future time to raise its prices and decide on what quantities of kerosene it wished to market. The government of India's main anxiety was still to avoid politically damaging shortages and hence a clamour for rationing.

In Burma the authorities understandably had graver objections. There the company's market was largely insulated from foreign competition and it could provide almost all requirements at the agreed maximum price. To continue that would have given the Burmese consumer the right to buy kerosene at only half the price applying in India, thus involving a risk of large-scale transhipment. Hence Burmah Oil and Asiatic were forced to lay down a uniform price for the whole Indian empire. The lieutenant-governor protested with vehemence but unavailingly against what Ritchie feelingly termed "just another case of the Burman being made to pay for his Indian brother". Already rice cost far more than in wartime, and excessively high prices for two of the basic commodities were quite inopportune now that political groups in Burma were becoming more articulate and restive under British rule.

Despite all snags, the pool agreement came into effect on 1 June 1919. Watson later claimed to have forced it on Asiatic, which thereby lost the right to purchase Burmah Oil's surplus kerosene cheaply and sell it at current prices. In future it would simply receive a commission on those quantities of the company's oil it happened to sell. Although Burmese consumers were hit, those in India as a whole benefited from 1919 to 1924 by £6,400,000 a year on average – according to Watson's calculations: these were inevitably speculative since no one could judge whence and at what price kerosene would have come to India had the indigenous production not existed.

Watson reckoned that Burmah Oil itself gained little financially out of the agreement, but no costings have survived to check his claim. On the credit side, it was at last free of the old restrictions on output. He went on to negotiate a new petrol agreement with Asiatic, which gave him a better price for the increasing quantities of petrol expected from the improved refining methods.

By then the air age had come to Burma, as elsewhere in the world. During the first-ever flight from England to Australia in 1919 the company supplied the pilot (Sir) Ross Smith with fuel. A rival plane, fuelled with Indo–Burma petrol, came to grief: Watson declined to draw from that mishap any lessons about the relative properties of the two brands of fuel. Nor did he see any possible benefit to the company from buying the remains of the wrecked plane and putting it on show as an awful warning against relying on Indo–Burma's products, as Ritchie had playfully suggested.

Contacts with British Burmah were harmonious if unexciting. It was the largest rival company in Burma, with crude output of about a tenth of Burmah Oil's, but with none too robust finances that were over-dependent on loans. Even so, it fully honoured its product agreements with the company. Indo–Burma, on the other hand, remained a thorn in Burmah Oil's side, with its own distribution system and very adroit marketing methods. But Jamal still hankered after selling his half share to Burmah Oil, and tried out the idea on Cargill during the latter's Rangoon visit in 1920–1. He offered his shares at five times their nominal value: in Cargill's private opinion "an absolutely ridiculous proposition", although he asked Jamal for documentary evidence of Indo–Burma's current performance and prospects. The best Jamal could do was to produce a rather outdated report. Cargill and Ritchie then told him bluntly that they were not interested.

They would have been interested enough if they could have gained outright control. However, that was out of the question as long as the other 50 per cent interest was held by Steel Bros, which had consistently warned Jamal against having any truck with Burmah Oil. Yet one of their employees had confidentially admitted to Ritchie that Jamal was "in a hopeless mess financially and that they were having a great deal of trouble with him." Providence seemed to be on the arch-gambler's side when Indo–Burma made an oil find at Lanywa, as mentioned in Chapter XVII, for he had been speculating so heavily in rice and other commodities that this windfall temporarily helped to shore up his ailing empire.

After fruitlessly appealing to Ritchie again, late in 1921 Jamal had to suspend payments. Indo–Burma was unaffected as his half share was disposed of elsewhere, but at a stroke he forfeited the commercial standing he had hitherto held in the province. "I have no sympathy for Jamal personally," Ritchie wrote home, "but it is rather sad that a man of his undoubted ability . . . who has done more for the province than any other Indian or Burman should come down through sheer greed." Jamal never really got over this collapse and died in 1924.

Indo–Burma did not scandalize the Burmah Oil directors in the same way as some newly formed enterprises did. The Indo–Burma Oilfields (1920) Ltd, for instance, put out a prospectus that Cargill and Watson claimed to be full of blatant lies. Ironically, when it was on the brink of collapse in 1923, its management appealed to Watson; he refused to help, and it went into receivership. The Anglo–Burma Oil Company had an even more dismal fate. Cargill and Greenway had branded it as a "ramp" on its establishment in 1910, and it had succeeded in extracting from gullible British investors no less than £500,000, all of which "melted away" without helping to produce any oil. In 1922, irregularities led to dealings in its shares on the London stock exchange being suspended, and two years later it was in liquidation. Watson reckoned that the goings-on by this pair of companies did "more harm to the credit of the province than all the failures which have taken place in Burma during the past decade": and that was saying a great deal.

Relations between Burmah Oil and Anglo–Persian throughout this period remained as delicate as they had been in the wartime years. The splitting of the companies' day-to-day

operations had been of far more than merely symbolic importance. As early as February 1918 Watson had had to remind
Cargill that Burmah Oil might in the more distant future have
to look upon Anglo–Persian as merely a source of income, with
no automatic say over its production or marketing policies. He
foresaw that Anglo–Persian would eventually be pursuing
policies incompatible with those of Burmah Oil, so that a
different team in charge there could well "take an active line of
their own irrespective of any Burmah Oil interests". He was in
the meantime concerned with two major questions affecting
that company: how capital could be raised for its growing
financial needs, and how to avoid the possibility that the future
flow of its products might disrupt the Indian market.

On finance, since the government was a sleeping partner,
Burmah Oil had management responsibilities as the only other
ordinary shareholder of note. Now that the war was over,
Anglo–Persian embarked on an ambitious programme of
world-wide expansion and Watson, as a director, was not
backward in contributing his views on the expansion programme and other matters. It had already raised a further
£2,800,000 in preference shares and debentures but to maintain
a financial balance a new ordinary share issue was imperative.
At the end of 1919 the government and Burmah Oil agreed to
take up their portions of a £4,500,000 issue, only part being
called up at that time.

Meanwhile, Anglo–Persian had managerial problems to
overcome. In 1919 Greenway was reinforced with no less than
five managing directors, including Garrow and Lloyd. Yet he
still found himself involved in far too much day-to-day work:
hence in mid-1921 Cargill and Watson had a wide-ranging
discussion with him in Glasgow – significantly, not London –
on his company's organization. Clearly they met him not as
paymasters but as ordinary directors with specialized knowledge. As a result he more or less successfully freed himself from
the routine to concentrate on long-term issues.[1]

To help bolster up the still inadequate senior management in
Anglo–Persian, Hubert Heath Eves, general manager in India,
came home that year. When in 1924 Garrow's health broke
down and he resigned as director, Eves took his place. Sadly,
the talented but over-sensitive Garrow died of alcoholism in
1931 when he was only 53.

Anglo–Persian had further financial troubles in mid-1921, when the Treasury refused at a time of retrenchment to pay up its call on the ordinary shares issued in 1919. Cargill and Watson personally appealed to the Chancellor, Sir Robert Horne (see Chapter XIX), but in vain. His refusal automatically restricted the amount of outside money the company could raise; not until a year later did parliament finally release the sums due. Meanwhile Watson was striving to avoid future difficulties by drafting a document that would clarify Anglo–Persian's relations with the government.

He proposed that Whitehall should allow it to raise further ordinary share capital from outside sources, subject to maintaining government control, but at well below the then existing two-thirds shareholding majority. It would moreover define the circumstances in which government directors could use their powers of veto. Watson's package went sour because at the same time he sought from ministers a clear declaration that they would give Burmah Oil the first refusal should they ever decide to sell their Anglo–Persian holdings. He chose to believe that that had been implicit in the 1914 agreement: for him it was a vital aspect of the merger proposals he was pressing at a very high level (see Chapter XIX). Whitehall declined to make such a declaration, and he never achieved the accord for which he had hoped.

Paradoxically, Whitehall was not averse from interfering when it thought necessary, as when in the autumn of 1922 Stanley Baldwin, President of the Board of Trade, proposed the appointment of Sir John Cadman as a managing director of Anglo–Persian. By then Greenway was 65 and thus within sight of retirement. Since there was no heir apparent within the company, the Board of Trade and the Petroleum Department (the former Petroleum Executive) were undoubtedly seeking to have Cadman groomed as the eventual successor to Greenway.

Cargill and Watson both reacted furiously to Baldwin's initiative as going far beyond the government powers laid down in 1914: whether they had an alternative candidate of their own is not recorded.[2] Certainly Watson, who might have made an interesting chairman of Anglo–Persian, was fully committed to Burmah Oil. He may in any case have been too much identified with the Shell group to be fully acceptable. In a bizarre compromise, the whole matter was dropped for six months whereupon the Anglo–Persian board, off its own bat, appointed Cadman as managing director.[3] Further ructions were to occur between the government and the Cargill–Watson team before Cadman was finally nominated to succeed Greenway as chairman in 1927.[4]

The problem remained as to what would happen when the agreements with Asiatic ran out in 1922 and large quantities of Anglo–Persian products began to appear in the Indian market. Those products were of foreign origin and therefore subject to the tariff, despite earlier pleas with Delhi to have them exempted. Greenway had had the bright idea of requesting the Foreign Office to induce the government of India to acquire Abadan and neighbouring territories by outright annexation or purchase, but that idea got nowhere. Instead, in July 1921 the two companies agreed that neither should seek to penetrate the other's established markets. As Watson candidly explained to Ritchie, that was simply a device for excluding Anglo–Persian from India and thereby helping them to ensure the maximum market share for Burmah Oil's products. Yet that could only be a holding operation, and far more fundamental changes in India had to be delayed until a permanent marketing organization was eventually set up in 1928.

At that time Watson was much in demand as Anglo–Persian's trouble-shooter. In 1920 he went to Romania and brought the oil company Steaua Romana under British control, a substantial stake being held by Anglo–Persian. Burmah Oil held no shares, but supplied the resident managing director who was an ex-Finlay Fleming & Co. man.[5] At home, in 1919 Watson took the initiative to save the ailing Scottish shale-oil companies. Although they had been hit by the competition of mineral oil, the government had refused to prop them up with subsidies. Their leader and spokesman, William Fraser, had already organized some self-help by setting up the Scottish Oil Agency Ltd to market their products jointly, and now enlisted Watson's help. By September 1919 Watson had been instrumental in forming Scottish Oils Ltd, a combine of the main shale-oil producers, under the control of Anglo–Persian. The latter guaranteed the running of the shale-oil refineries at full capacity by supplying such extra crude as was required.

Fraser was appointed managing director and Watson an ordinary director; with two such pundits on Scottish Oil's board, Cargill as a prestige chairman had little to do. By a side-posting reminiscent of Wallace's recuitment to Burmah Oil, Fraser joined the Anglo–Persian board in 1923 and became a managing director in 1928. During the 1930s and 1940s he, Watson and Frederick Godber as managing director of Shell were the three acknowledged executive heads of the oil industry in Britain.

The difficulties of finding reliable supplies of oil in Burma or India caused Burmah Oil to look into propositions elsewhere. Some in Venezuela and Thailand, for instance, were considered and rejected: however, in 1921 the directors agreed to join Anglo–Persian in a Hungarian oil syndicate. Whigham was the syndicate's chairman. Cargill, as ever, shrank from involvement in unfamiliar parts of the world, which seemed to bring ever more worry with scant chances of profits. Yet he was very susceptible to pressure by his colleagues. In 1909, when offered a concession in Trinidad, he had declined, stating that the island was "really quite outside our sphere". Yet only three years later Burmah Oil took a quarter share, together with the Shell group, in the expansively named United British West Indies Petroleum Syndicate Ltd. Ironically enough, Cargill had been "forced" into it partly by Redwood and partly by the anti-Shell Greenway, director of a trust company that was the third party in the syndicate. The fourth quarter-share was held by the Paris Rothschilds.

    The outcome of operations was as disheartening as Cargill had feared and by 1918 he was all for pulling out. But by then he had the tenacious Watson to contend with, and Watson persuaded him that it would look very bad to sever connections there while seeking a merger with the Shell group as a whole. The hapless Cargill had to see another £150,000 of good Burmah Oil money being thrown after bad, for the venture failed to earn any dividends. By 1923 the syndicate shares were written down to a nominal £1 in the Burmah Oil books: the clearest of indications that the company's involvement in Trinidad was nearing its end.

    Watson's lively suspicions of Standard Oil, so freely expressed to Glasgow in the final months of the war, were not dispelled when peace returned. Yet even he could not have foreseen how Anglo–American relations would be soured by a bitter political fracas over oil. By then the United States was a net petroleum importer, and its businessmen and politicians vociferously demanded an "open door" policy to give all nations equal opportunities to explore for oil anywhere in the world. As British interests appeared to be making supreme efforts to grasp an unduly large share of the world's reserves, most notably in Persia and Mesopotamia, American hostility was unrestrained. Relations reached their nadir in 1920 when Standard Oil, having been refused permission to prospect anywhere in Mesopotamia, denounced Britain for its dog-in-the-manger attitude.

In the subsequent war of words Burmah Oil was not spared, the State Department in Washington reporting as follows:

> American oil companies are expressly excluded from doing business in Burma by proclamation signed by Queen Victoria and Lord Salisbury, Secretary of State for India, on September 24th 1884, and a blanket concession of ninety nine years was given the Burma Oil Company (Limited) on August 23rd 1885 protecting this company from all foreign competition.

The India Office at once dismissed the alleged corroborative documents as "self-evident forgeries". Salisbury was not in office during 1884, Queen Victoria issued no proclamation on the subject, and Burmah Oil did not even exist until 1886. Indeed, a blanket concession of that kind would have surpassed the directors' wildest dreams and of course was never granted. The State Department later amended the date to 1889, but no documents of even the remotest relevance have been located in the British public archives.[6] Relations with the United States very gradually improved and by 1922 Watson was beginning to speak directly with Standard Oil executives, although not to any great effect.

What most irked the Americans by then was the projected merger between Burmah Oil and Shell. Late in 1921 Ritchie reported to London, apparently on the strength of information gleaned in Calcutta, that they would by every possible means prevent the formation of such a combine, but Watson brushed aside such fears.

As it happened, A. C. Bedford, the chairman of Standard Oil, had that November been to see Watson as Anglo–Persian's troubleshooter about something to do with Romania. In an aside, Bedford floated the idea of "some form of co-operation", even an amalgamation with Anglo–Persian, clearly as an alternative to Shell's merger proposals. Watson turned the tables on him by enquiring, "Would or could you contemplate an arrangement which gave British control over the Standard?" Bedford angrily rejected the idea as quite unthinkable, and spoke of the two countries' common heritage, in contrast with the foreignness of Shell: sentiments that Watson impatiently brushed aside as totally irrelevant. "Neither of us has at the moment anything tangible to put forward to the other," he concluded bluntly. He did, however, drop a hint about a possible market-sharing arrangement; alternatively, he suggested to Bedford that Standard Oil should by rights withdraw from India as its remotest and least profitable market.

The two men met again in March 1922 when Bedford was in London pressing for a share in the Mesopotamian concession.[7] Once again he suggested a merger between Standard Oil and Anglo–Persian. Watson gave no quarter, demanding that any terms from America should be at least as attractive for Burmah Oil as those already offered in its merger scheme by Shell. Bedford could only propose a conference of all interested parties in New York; meanwhile he and the American ambassador in London would be lobbying ministers in Whitehall, presumably to have any decisions on the Shell scheme postponed. Bedford worked also on Greenway, who repeated to Watson some of the American's observations, notably that:

> it would, in his opinion, be a disaster that the two great English-speaking nations should be divided on a question as important as this, in regard to which a too hasty decision might seriously endanger the future relations between the two countries.

Watson's written comment on the letter was crude but to the point: "balls!"

No more was heard from Bedford about a possible merger. Instead, Watson became keen on trying to arrange a marketing agreement with Standard Oil, for that company was undercutting kerosene prices in Burma apparently as part of a move to increase its market share throughout India. This time Watson wrote to W. C. Teagle, who as a young man had helped to resolve the rate war of 1908 and had since become president – that is, managing director – of Standard Oil. As sales of American kerosene in Burma were not large, Burmah Oil was willing to provide the quantities required for sale under the Standard Oil label. Bedford and Teagle sought to entice him over for a conference in the United States, but he declined on the familiar grounds on having more important things to do.

Teagle therefore sent over an executive, C. F. Meyer – the peacemaker of 1911–12 (see Chapter XI) – to arrange an accord in India and Burma. Meyer was both antagonistic towards Shell and an exceptionally tough bargainer. He refused to have his company's kerosene sold through Burmah Oil outlets or let Burmah Oil dictate prices. However, as a gesture of goodwill he did discontinue the earlier underselling that had caused so much bad blood; thus something was achieved by his visit, and relations between the two companies remained fairly amicable for several years.

Meanwhile Watson, in Burmah Oil's name, helped to form several new enterprises which reflected some of the novel conditions in the post-war era. The Tinplate Company of India has been mentioned in Chapter XVII, but the candle-making industry in Britain was in a poor state and some kind of rescue operation was imperative if it were not to go under altogether. The leading firm was still Price's Patent Candle Company, which had refined Burmese oil in the 1850s and had subsequently been a purchaser of the company's wax until American rivals had offered cheaper supplies.

Whether there had been any regular trading between the two companies in the twentieth century is doubtful but, since 1919, Price's as well as several other candle companies had come under the ownership of Lever Bros presumably because of the soap and glycerine connections. It is not known why or how negotiations got under way, but by the spring of 1921 Watson was writing lengthy memoranda and having private meetings with Cohen and Lord Leverhulme, the septuagenarian founder and chairman of Lever Bros.

As often happened, discussions were very protracted, and not until early 1922 did Burmah Oil, Asiatic, Scottish Oils and Lever Bros reach agreement. As a result, they jointly formed Candles Ltd to buy Price's and the smaller candle firms from

Lever Bros and also take over Burmah Oil's candle-making plant in Rangoon. The share capital was fairly modest, but Burmah Oil found itself having to acquire preference shares and make loans for the long overdue refurbishing of Price's factories. By the end of 1924 it had about £500,000 sunk in Candles Ltd. Unlike many such ventures, that company survived. In 1982 Burmah Oil sold its 36 per cent of the equity to Shell.

Not unexpectedly, the last word on external relationships came from Asiatic. In the closing months of 1924 Watson was in India and Burma, while Cohen was coincidentally touring the world to inspect Shell's overseas installations. That December they therefore met for a joint session with their local representatives in Calcutta: typically enough, this proved a prelude to an extraordinary tiff about some minor agreement that Cohen condemned as unacceptable. Watson counter-attacked savagely, adding:

> If generally you are tired of our relations they can easily be terminated. Neither of us is dependent on the other; we are both "big boys" and can quite well, if necessary, walk alone.

There was much more in the same vein and it was enough to bring Cohen to heel. He assured Watson the links they had devised between their companies were "of enormous value to both sides so long as we continue as hitherto to discuss absolutely frankly and apply with great energy necessary measures to keep them always up-to-date and a source of maximum strength and efficiency to both concerns." In fact, there was no fear that either would be at all backward in speaking his mind to the other.

However much he might pummel Cohen, Watson recognized that the two enjoyed "what amounts to in effect a completely mutual partnership" on the marketing side in India. So he proceeded to write a joint memorandum of guidance for both their local agencies. That stressed the need to promote demand and public goodwill, to capture trade from outside companies and to pursue improvements in the efficiency of the two agencies as well as economy in distribution and marketing. He went on with an intriguing gaze into the future.

"In the changing economic conditions of India, experiments in these directions [of greater efficiency] may frequently be necessary."

Probably no one took the hint at the time; yet he had already agreed with Cargill, before their departure for the east, that Finlay Fleming & Co. would in the fullness of time cease to be the managing agency and Burmah Oil would instead establish a branch office in Rangoon.[8] That change took place in 1928, when the newly formed Burmah–Shell Oil Storage and Distributing Co. of India Ltd provided the inter-company trading partnership to which Watson and Cohen had aspired for so long. The same year they set up a joint Burmah–Shell office in Calcutta, thereby breaking the 60-year-old link with Shaw Wallace & Co. and its predecessor agency. There were to be far-reaching consequences for the company's internal performance from that historic step.

Notes

1 Ferrier, "Early Management Organization of British Petroleum", pp. 136–7.
2 It is almost certain that their candidate was J. D. Stewart, a Glasgow shipping magnate and managing director in charge of sea transport since 1919, but apparently he did not carry the confidence of enough directors. See Ferrier, ibid., especially p. 138.
3 Some of the correspondence is in PRO POWE 33/96. Greenway's letter to Cargill of 12 September 1922, suggesting that it would be better to give in rather than have Cadman forced on Anglo–Persian in face of board opposition, was taken by Cargill as evidence that Greenway was out to support Cadman, "and I'm afraid we will have very great difficulty in defeating him *this* time!"
4 After some heated correspondence between Churchill and Cargill in August 1925, Watson suggested that Cargill himself should be put forward for the chairmanship: letter Watson to Cargill, 21 September 1925. Cargill wisely declined.
5 For Romania at this time see M. Pearton, *Oil and the Romanian State* (1971).
6 US Department of State, *Foreign Relations of the United States 1921* II (1921), pp. 77–80, cf. J. Ise, *The United States Oil Policy* (1926), pp. 456–7. Among the sources of this over-researched topic are J. A. DeNovo, "The Movement for an Aggressive American Oil Policy Abroad 1918–20", *American Historical Review*, LXI, (1955–6), pp. 854–76; G. D. Nash, *United States Oil Policy 1890–1964* (1968), ch. 3, and M. J. Hogan, "Informal Entente: Public Policy and Private Management in Anglo–American Petroleum Affairs 1918–1924", *Business History Review*, XLVIII, (1974), pp. 187–205.

7 The efforts of Standard Oil in those years are dealt with in G. S. Gibb and
E. H. Knowlton, *History of Standard Oil Company (New Jersey): II The
Resurgent Years 1911–1927* (1956), ch. 11.
8 AGM 1928, *Glasgow Herald,* 12 June 1928.

CHAPTER XIX

The Merger that Never Was: (2) A Fight to the Finish
1920–4

During 1920 and the first half of 1921, as Whitehall sought to cope with one post-war problem after the other, little considera-tion was given to the merger question discussed in Chapter XVI. Now that the issue of Mesopotamia was deadlocked after the collapse of the Harcourt–Deterding agreement, the advan-tages from turning the Shell group into a British company seemed far less obvious.

Then Greenway quite unexpectedly jerked the dormant issue back to life.[1] Any chances of resurrecting Cowdray's all-British merger proposals had finally vanished when in 1919 the Shell group took over Mexican Eagle and established Shell–Mex Ltd as a joint marketing organization in Britain. In July 1921 Greenway informed the Board of Trade that Shell–Mex had asked Anglo–Persian to join a common marketing arrangement for petrol. This, it was argued, would economize on the capital cost of installing petrol pumps which were at that time begin-ning to replace the individual cans hitherto used by garages, and would thus help to keep down petrol prices.

Stanley Baldwin, president of the Board of Trade, and his junior minister in charge of oil, Sir Philip Lloyd-Greame, called in Cargill and Greenway for a discussion of the Shell–Mex proposals. Lloyd-Greame resolutely opposed any marketing link-up with the Shell group as a step towards monopoly; for some years that group in particular had been in bad odour for allegedly profiteering from high petrol prices. Cargill reckoned that the joint marketing proposal would save Anglo–Persian about £2,500,000 in installation costs, at a time when the government was refusing to supply it with further capital (see Chapter XVIII) and therefore hindering its ability to secure extra funds from the general public. Indeed, he and Watson had just been to see the Chancellor of the Exchequer, Sir Robert Horne, but without persuading him to change his mind.

Baldwin's own comment from the chair is instructive. He felt the proposal was perfectly sound from the business point of view, but had political snags, being vulnerable to attack in parliament and from the general public. As a former iron-master, he was well acquainted with trade agreements and company mergers and saw little harm in them as such: this personal opinion, which he later extended to the merger scheme, was – in Sherlock Holmes's words – the one fixed point in a changing age. As over the next few years he moved up the political ladder to the prime ministership, his benign attitude gave great solace to the proponents of the merger. Among those proponents could now be numbered Cargill, for Watson had been working on him and had converted his former timorous and lukewarm attitude into one of modified enthusiasm. When, at the meeting, Lloyd-Greame declared that the corollary of the fund-raising problems was that the government should "clear out of" Anglo–Persian entirely, Cargill was quick to reply that the Harcourt scheme of 1916 should be looked at again.

Never one to let an opportunity slip, Watson followed this up with a memorandum to Baldwin setting out the advantages of such a merger.[2] Effective control, he said, should be achieved by bringing in Anglo–Persian, now essential as a counterweight because Royal Dutch had since 1916 grown faster than its intended British partners. Britain would thereby secure a majority of 50.2 per cent to the Dutch 49.8 per cent. That scheme would automatically take in the more restricted petrol marketing proposal.

While ministers and officials were digesting his memorandum, Watson discussed it with Sir Louis Kershaw, head of the industries department at the India Office. Watson made it plain that there was now to be a fight to the finish over the merger scheme: it must either be accepted all round or be definitely vetoed by the government. The unenthusiastic Kershaw raised two objections: first, the fear that the new combine might decide to "sit on" Indian or Burmese concessions without working them, thereby depriving the local governments of revenue, and second, the possibility of competition being entirely eliminated, with a consequent "bleeding" of the Indian consumer. Watson's response was that Delhi had full sovereign powers to curb such harmful practices, if necessary by legislation. Kershaw was curious to know why Greenway had so unexpectedly become a friend of Shell, but Watson was not to be drawn on such a sensitive question.

Other government departments had their own reservations over Watson's memorandum. When Lloyd-Greame convened a meeting of officials, the Admiralty representative emphasized the possible threat to cheap fuel oil supplies from a stifling of competition in both India and Persia, while his opposite number from the Board of Trade argued that effective and total control over Royal Dutch and Shell would be worth paying for, but only provided that it was not the consumer who had to pay. These reservations led all present to agree that Watson should be asked to give more precise views on the two really fundamental issues: how to safeguard consumers' interests, and how to secure a lasting and really watertight British control over the proposed combine.

Within a few days Watson had submitted fresh memoranda.[3] On the first question, he pointed to the millions of rupees customers in India had been saved from paying as a result of Burmah Oil's maximum price policy for kerosene. Some form of legislated profit limitation might be an additional safeguard, but he would much prefer the government to leave matters to "the ordinary laws of supply and demand". Unless a virtual monopoly were to occur, the "artificial" – presumably government-imposed – control of prices and supplies would in his opinion be both dangerous and economically unsound. Watson went on:

> If petroleum companies have no soul they at any rate have some modicum of business acumen, and are therefore not only fairly sensitive to public opinion but well alive to the fact that high prices retard and low prices stimulate the demand for their products.

There he echoed the views of Deterding and Cohen, both of whom stressed the vital role of supply and demand in keeping markets properly serviced: he neglected the fact that where no alternative supplies were available, collusion and monopolistic discrimination in markets could greatly harm consumers, who in practice all too rarely benefited from self-denying acts such as Burmah Oil's maximum price policy.

Watson's second memorandum, on the question of achieving permanent and effective British control, did little to allay officials' disquiet because he seemed to be admitting that a good deal would have to be taken on trust. Practically all of Britain's oil supplies came from overseas, he conceded, and he hoped that British supervision of any merger would be as watertight as possible. Yet he dodged the paramount issue that had preoccupied the Harcourt committee: how to bring Shell firmly and irreversibly into the British camp. The sceptical J. C. Clarke, Cadman's successor as head of the Petroleum Executive (now the Petroleum Department), therefore dismissed his arguments as "not very successful".

At a higher level, ministers were interested enough to keep in touch with Cargill on this topic, Horne and Lloyd-Greame having an hour-long meeting over tea with him. Fearful that his remarks had merely awakened the doubts of two key ministers, Cargill urged Watson to see Lloyd-Greame; since he could deploy his case to greater effect orally than through his rather convoluted prose, Watson at once took up that suggestion. To the minister he stressed the need for haste as so much had to be done and the sooner things were under way, the better. Lloyd-Greame neatly side-stepped this one: nothing, he said, could be decided until the international naval conference just opening in Washington was concluded. If the merger scheme went through before then, Standard Oil, still a very powerful pressure-group in the American capital, might seek to create difficulties over any decisions on naval disarmament agreed by the conference.

He did not fall for Watson's enticing suggestion that the government could always privately accept the scheme in principle but defer a public announcement until the conference was over. He also emphasized that the government of India must be consulted; Watson, still bruised by his discouraging encounter with Kershaw, privately dismissed that as a time-wasting and therefore futile exercise since Delhi was virtually a puppet on Whitehall's string.

Even so, Lloyd-Greame was sufficiently impressed by Watson's arguments to compose a lengthy cabinet paper, proposing a merger along the 1916 lines but to include Anglo–Persian.[4] In January 1922 it was submitted to the cabinet, which appointed a committee, headed by Stanley Baldwin, to go into the scheme in full detail. The other members were all representatives of departments such as the Treasury, Colonial Office, Admiralty and Petroleum Department.

Meanwhile Deterding, already under deep suspicion in Whitehall for his failure to deliver Shell into British hands in return for an enlarged stake in Mesopotamia, did his cause no good by adopting some shock tactics designed to force a rapid decision. Through Watson he claimed that he had been summoned to an important top-level conference at The Hague with the Dutch council of ministers in order to settle various controversial issues between Royal Dutch–Shell and the government there. He must take over with him the British government's consent to the scheme since – he asserted – his Royal Dutch colleagues' earlier approval in principle had been made only out of pique because the council of ministers had shown so little concern for Dutch oil interests. Watson commented that although it all sounded a little too much like a fairy story, he personally believed that Deterding was "honest in the matter". In fact, nothing more was heard of it.

To the Baldwin committee, as to earlier ones, the opinions of Burmah Oil's directors would be a vital factor, and in March Cargill and Watson were called to give evidence. As chairman, Baldwin sought to draw them out on what could well become a sticking point: how British control could be safeguarded. Watson, who did most of the talking, reckoned that a merger to include Anglo–Persian would not require elaborate new machinery. Instead, it could be put through on the basis of the various units' existing capital values, but with the creation of special voting shares to give a right of veto over any share transfers calculated to nullify such control.

Committee members had already interviewed Greenway who, while claiming to be all for the merger in principle, had in practice been less than enthusiastic. They therefore put to

Watson an apparently innocent question: was Anglo–Persian reluctant to join a combine through fear of damaging its bright prospects for growth? Watson played down that company's optimistic ideas about the future and judged that it would get a perfectly fair deal out of a merger on the proposed terms. Since both Burmah Oil and Anglo–Persian had to rely for their supplies on no more than a few fields, he went on, they should find a merger attractive because it would spread risks world-wide. Whether or not he had convinced the committee, he had made the best of his case, and must now await its findings.

Then disquieting reports reached the company via Standard Oil's London subsidiary that Greenway was trying to get the American administration to defeat the merger scheme by diplomatic means. Those reports angered Cargill, particularly when a calculated press leak led him to believe that Cadman, a long-time trusted adviser in Whitehall who had joined Anglo–Persian as a technical consultant, was now equally antagonistic to the scheme. It remained to be seen whether Cohen, after giving evidence to the committee, was right in thinking that its members fully supported the merger in principle and were merely taking their time over consequential points such as how and to whom the government might sell its Anglo–Persian shares.

In fact it was the disposal of those shares that finally made the committee decide that the disadvantages of a sale out-weighed the benefits.[5] In July 1922, after Baldwin had submit-ted the negative report of his committee, the cabinet decreed that Anglo–Persian should not be included in any merger.[6] Baldwin was considerate enough to write a personal letter to Cargill, who would personally have accepted the decision, but Watson was determined to fight on. He asked Baldwin to give him the precise reasons for the outcome, explaining that the Burmah Oil directors wished to consider whether to continue merger talks with Royal Dutch–Shell. This was rather diseng-enuous, as Burmah Oil's ordinary share capital would cover less than 13 per cent of the gap between Royal Dutch's 60 per cent and (British) Shell's 40 per cent, thus providing no possibility of achieving British control.

In September Baldwin received Watson, and lived up to his later reputation of speaking with appalling frankness. As an industrialist, he said, he personally commended the scheme,

which was a sound commercial one; moreover, he disapproved
of government either having an interest or presuming to interfere in business. But the cabinet had had to decide on purely
political grounds, such as the possibility of an alliance between
Pretyman, who maintained a fatherly interest in the success of
the Persian venture he had helped to promote, like-minded
Conservatives, and Labour members who would fight any sale
to private interests of profitable and strategically vital public
assets. This alliance would be reinforced by ordinary members
and electors of all parties whose body was "the trust" (or
monopoly) in any form. Government simply could not face
such a lobby: it was asking for trouble. Hostile reaction could be
expected from the United States as well. Hence for ministers
"the situation was there and they had to deal with it as they
found it."

Even so, Watson derived much comfort from Baldwin's
assertion that political reasons alone had led to the rejection of
the scheme, despite its commercial advantages. He therefore
decided to take the smaller merger – omitting Anglo–Persian –
back to the authorities in Delhi via the India Office. However,
Kershaw proved to be even less enthusiastic than he had been
in 1921 over the large merger, and simply referred the idea to
India. Watson, by then buoyed up with entirely irrational
optimism, heard from Kershaw in December of Delhi's decision that the latest scheme was totally unacceptable.

For once in his life Watson took the news very badly. Like a
mortified schoolboy he replied, "So far as I personally am
concerned this decision, one way or another, on either proposal
is of course a matter of indifference: I am possibly fortunate
enough not to take these business matters too personally."
Kershaw was thoughtful enough to send Watson a private
letter, which merely repeated familiar arguments as seen from
the other side of the world: that amalgamation might lead to the
creation of monopoly to which powerful foreign interests would
be admitted, that smaller oil companies might be squeezed out
of existence, that no benefit would result to the Indian consumer and that the measure would be intensely unpopular in India.

Watson was now in a very tight corner. However, wishing to
keep the scheme in their minds, he sent copies of the correspondence to Baldwin, who had lately become Chancellor of the
Exchequer, asking that Lloyd-Greame, his replacement at the

Board of Trade, should see them as well. Yet, whether Watson realized it or not, the Baldwin committee's thumbs-down of July 1922 effectively destroyed the final hope of any merger between Burmah Oil and the Shell group.

The events of the following two and a half years are no more than a pointless epilogue, in which too many people spent too much time in striving to breathe life into a scheme already past resuscitation. It only heightened his wounded feelings to hear from Shell in February 1923, via Slade who was still non-executive vice-chairman of Anglo–Persian, that Greenway was "in favour of and anxious" for a full merger, to include his own company. However, he had reportedly been "somewhat annoyed at and jealous of the Burmah Oil Company" for having started up the post-1920 round of negotiations. Hence he had dragged his feet during those negotiations.

Watson had every reason to be sceptical about a piece of hearsay and could not decide whether Slade, whose intellectual gifts were not regarded as being of the highest, was merely "terminologically inexact" (a phrase borrowed from Winston Churchill) or whether he might have misunderstood what Greenway had actually said. "The *volte face* is too much to swallow," Watson told Cargill. Instead, he urged his Shell friends in staccato phrases that for once had a Kirkman Finlay-like ring to them, "go on, probe the matter further . . .; commit him [Greenway] up to the neck – and then we'll see what we'll see!" Yet there appeared after all to be some fire amidst the smoke when in March he learnt to his astonishment that Greenway was to dine with Sir Marcus's son Walter Samuel, now chairman of Shell, and with Cohen.

To be sure, Greenway's personal support or antipathy were of little consequence as long as cabinet policy was to keep Anglo–Persian out of any merger scheme. Then in July 1923 Cohen had a brainwave. He decided to enlist the aid of a top political figure: none other than Winston Churchill, out of office and out of the House of Commons since his election defeat of 1922. He was a crucial link for a number of reasons. He had in 1914 made the government a majority shareholder in Anglo–Persian; to Watson he was "probably the most powerful and dangerous critic" of the recent merger scheme; and perhaps most significantly, he had an easy relationship with Baldwin, now prime minister, and was therefore a good person to assuage Baldwin's political fears and get him publicly to back the scheme which he had privately commended in principle.

Edwin Montagu, former Secretary of State for India and a friend of Churchill, happened to be Cohen's cousin, and he agreed to arrange a private dinner party to include Churchill. There Cohen dangled before the great man the succulent bait – skilfully concocted by Watson – that a merger would place the "coping stone" on his achievement of 1914 in securing the government stake in Anglo–Persian, by clinching British control over a world-wide oil producing, marketing and transport network.

Cohen later gave a full report to Watson, who passed it on to Cargill:

> Winston at once saw the picture complete and volunteered the explanation that when last put up [in 1921] he and his colleagues of weight had been too immersed in more urgent and important matters of state to give the proposition any sort of consideration, or to attempt to understand it as he now clearly did.

Then Cohen asked him the direct question: was he "prepared to take it up and endeavour to put it through?" He needed time, Churchill replied, to consider all sides of the matter: whether the scheme was genuinely in Britain's interests; whether the government as a whole could be brought to a change of heart after the Baldwin committee's flat rejection; and whether becoming personally involved might adversely affect his own career, which had always been a political one and which he intended should remain so.

He did explain, however, that the lack of a ministerial income left him with his living to earn and he would need to lay aside the fourth volume of his major work on the great war, *The World Crisis*. Cohen readily agreed that a fee should be paid. Churchill then sought the advice of one or two former Admiralty colleagues and, after receiving encouragement there, agreed to go ahead.[7] He asked for written instructions from Burmah Oil and from Shell, so as to be able to say that his help had been expressly invited by the two companies. Watson therefore drafted a suitable letter for Cargill to sign.

Cohen next saw Churchill early in August 1923 and reported

that he was becoming "keener and keener", so much so that he would rather like to become a director of Shell, thereby breaking the rule of a lifetime to avoid all business commitments. That news disconcerted Cohen, who had cast Churchill in the role of honest broker and feared a disastrous weakening of Churchill's influence in Whitehall once he was seen to be an interested party. Before the idea went further, Churchill called at 10 Downing Street and saw Stanley Baldwin.

The first round seemed to be easily won when Baldwin showed himself quite prepared to look again at the merger question. He was "thoroughly in favour" of it "on the lines proposed"; namely, with safeguards for maintaining genuine British control of the new combine, and with terms satisfactory to both Anglo–Persian and Burmah Oil as its parent. He was much concerned about Anglo–Persian's unsatisfactory trading position lately, which had led to a decline in its profits by a third since 1920/1 and the halving of the dividend. The value of the government's holdings had therefore been seriously eroded, so that their sale would both remove a source of anxiety and bring the hard-pressed Treasury a welcome windfall of £20,000,000. As Churchill reported to his wife, Baldwin "might have been Waley Cohen from the way he talked".

By the end of the month Churchill had made up his mind that the best platform from which to carry the scheme through would be a seat on the Burmah Oil board rather than that of Shell. Watson was not at all keen on the prospect – which never came to anything – of having a very seasoned and argumentative politician as a fellow-director, but Cohen brushed that aside and composed a lengthy brief for Churchill.

After rehearsing the broad political advantages, Cohen also cited various commercial ones, although treating them as "all minor considerations". He concluded his brief to Churchill with a trowelful of old-fashioned flattery:

> If in undertaking this task you prolong for a little your withdrawal from the sphere of general politics and thereby render the nation poorer in one direction, I hope you may feel that you are enriching it to a more than compensating extent in another which is of no less vital importance.

Churchill took the brief and was soon devoting most of his working day to the matter, having preparatory talks with his crony the press baron Lord Beaverbrook among others. Yet as the prime minister was the key to the situation, Churchill was eagerly awaiting Baldwin's return from holiday. Meanwhile, to the companies, he quoted his fee as £10,000 if the deal did not go through and £50,000 if it did. Even Cohen was shaken by the magnitude of these sums, but Cargill quite sensibly observed that "we couldn't very well haggle or bargain with him" and proposed that Shell, Royal Dutch and Burmah Oil should share the cost.

On his return, Baldwin went to visit Churchill who had gone down with influenza, and reiterated his "full sympathy" with the scheme. Two departments above all remained to be won over, and he therefore advised Churchill to see Lloyd-Greame as President of the Board of Trade and L. S. Amery the First Lord. One other person Churchill felt he should work on was Greenway, but as Cargill reported to Watson – adding the question mark –

> W. C. rather funks (?) tackling Greenway and suggested Beaverbrook doing this. I said I thought C. G. would take it much better from W. C. I told Cohen . . . that you thought he [Greenway] was weakening in his opposition.

Burmah Oil duly paid up £5,000 on account to Churchill, which caused Cargill no end of worry as to how the payment might appear in the company's books for, unless the recipient's name were disclosed, the auditors would be unable to pass it. In the end, it was decided to ask the board to set up a secret service fund of £10,000 although, as will be shown, this was never needed.

As it happened, Cargill had his mind diverted from this anxiety, having just visited a medical consultant to make sure his heart had not been strained by bicycling and strenuous days out shooting. A check-up revealed that he was as robust as any man of his age, then 56. This gave him the gloomy satisfaction of being able to reply that there were plenty of other worse diseases than apoplexy he could die of.

Refreshed by a much-needed holiday, Churchill was soon pressing ahead with his negotiations. At the end of October 1923 he held a summit meeting of the interested parties, comprising Cargill, Watson, Greenway, Walter Samuel, and also Cohen who represented Royal Dutch while Deterding was in America. Together they agreed a memorandum, laying down mutually acceptable provisions for a merger along the lines of a British majority on the holding company's board, with each subsidiary maintaining its own identity.

Watson and Cohen then spent the best part of five days in very exhaustive discussions with Churchill, who was famed for mastering a complex brief by talking himself into it with the aid of experts. His house in Sussex Square must have rung with the voices of the three highly articulate and supremely self-confident men haranguing one another.[8] By then it was clear that the main target for the Churchill treatment must be the Admiralty: if and when their Lordships were won over, cabinet consent would be largely a formality.

Churchill therefore sent a memorandum to the Admiralty, containing assurances that the new combine would fully honour Anglo–Persian's vital fuel oil contract by means of a supplementary agreement with the government, which would secure the quantities involved on the combine's entire reserves. A Churchillian touch was that if at any time the First Lord should officially declare "reason of state", the combine would "discreetly assemble at any particular ports a quantity of oil equal to one year's Admiralty supply". Moreover:

> In the impossible event of any failure of the new combination to execute the supply contract in peace, emergency, or war the British government would be legally empowered to resume direction of the Anglo–Persian fields so far as might be necessary to secure the execution of the contract; and in addition of Sarawak and all other fields in British territory in the possession of the new group, all of which fields are within the physical control of British sea power.

Not to be outdone by Churchill in Napoleonic flourishes, Deterding – now returned from America – proposed that, should any threat to oil supplies occur:

a large number of tankers might during the emergency period be pushed into Tarakan harbour and filled at the convenience of the Admiralty, thus blocking entirely all other purchasers.

Tarakan island, where Royal Dutch produced oil that was suitable as naval fuel without having first to be refined, was in Dutch East Borneo. It was therefore specifically excluded from the merger provisions, and Deterding probably had no authority to make that proposal.

Churchill's memorandum went up through the Admiralty hierarchy to the Sea Lords, who discharged a return salvo of questions. One scored a hit below the waterline: about the fuel oil contract of 1914. There the terms over price and quantity were fully binding as long as the oil continued to flow from the Persian fields. Yet if for any reason those fields were to cease production, Anglo–Persian must continue to supply, having to buy in from outside at current market prices. In that case the Admiralty would be required to pay correspondingly more. The astonished Churchill professed never to have grasped that point earlier; persuasive advocates as Watson and Cohen were, their struggle to convince him was arduous and protracted.

Their Lordships raised some other contentious questions, for instance, would Anglo–Persian be allowed to keep open its refinery lately built at Llandarcy? Churchill replied that it had been losing money ever since coming on stream in 1922, so that if the Admiralty wanted its continuance for strategic reasons, the losses would at least partly have to be recouped in the price of fuel oil. That reply did not go down well in the Admiralty, where the general impression was that all his answers had been "entirely unsatisfactory from the point of view of naval interests".[9]

By then something had happened to Churchill himself. He was having second thoughts about the merger scheme. Perhaps he could already see that he would never succeed; however, more pertinently, a political crisis was beginning to blow up after Baldwin had inadvisedly raised the tariff issue. On 12 November a general election was announced, and Churchill received a number of approaches from constituencies with candidatures to offer. Three days later he informed Cohen that he was withdrawing from the negotiations. The £5,000 fee he had received eventually found its way back to Burmah Oil, thereby saving Cargill from the embarrassment of having to recommend to his board the secret service fund earlier contemplated.

Deprived of this powerful spokesman, the merger scheme began to languish. Watson tried hard to open a dialogue with Admiralty officials, but without much success. On 6 December the general election took place. Its outcome was that the Conservatives, while remaining the largest party, had no overall majority.

Watson at once wrote to Lloyd-Greame, requesting him to go ahead with the merger scheme, "involving the sale by the government and the acquisition by my company of the government's holding in the Anglo–Persian". That of course begged the still unresolved question of who should be allowed to buy that holding. The minister's realistic answer was that it must rest with whichever government took power once parliament had met. However, recognizing that the main outstanding questions were naval ones, he arranged for the First Lord, L. S. Amery, to attend an early meeting between the Admiralty and the company.

That was dismal news for Watson. Amery has been well described as "an ardent and devoted imperialist of the old school, and a strong believer in the beneficial influence of a powerful British navy".[10] He was therefore unlikely to gainsay the virtually unanimous views of his naval and civilian advisers alike. With the backing of the Admiralty board he submitted a cabinet paper arguing cogently against the merger scheme.

That paper systematically demolished his opponents' case.[11] As to naval fuel oil supplies, currently 40 per cent came from Anglo–Persian. Not only were the terms exceedingly favourable, but that absolutely guaranteed supply enabled the Admiralty to bargain very toughly with contractors for the remaining 60 per cent. Although the experts had dispelled the fear of the Persian oilfields going dry, the quaintly but aptly named *bon père de famille* clause in the combine's proposed document of guarantee, which contained a pledge not to allow the assets of subsidiaries to be neglected or their development stunted, in reality contained no really effective safeguards.

On grounds of relative cost, therefore, the Persian operations could well have to be curtailed to the advantage of, say, Sarawak or Borneo, which in distance and quality of product were far less suitable for supplying Britain. In short, Amery disliked the merger for two reasons. First, the future of the Persian fields and hence the fuel oil contract as it stood might be jeopardized; and, second, no offsetting advantages were being offered in the form of geographically convenient and wholly trustworthy alternative sources of supply.

On the projected combine, Amery's condemnation was a rerun of a very familiar charge. "It is no exaggeration to say that the group, together with the Standard Oil Company would practically control the oil fuel supply of the whole world," at the expense of British interests. Greenway must somehow have been getting at him, for he added that the management of Anglo–Persian had of late been seriously hampered by having the "threat of extinction" constantly hanging over its head. He pressed the government to reject the merger proposal outright, and declare the rejection to be final and irrevocable. At that point, Baldwin's administration was dismissed from office by a vote in the House of Commons, and Ramsay MacDonald formed a Labour government with Liberal support.

The first-ever meeting of a Labour cabinet at 10 Downing Street was a muted rather than an euphoric occasion. New ministers were so anxious not to antagonize city or international opinion that they had no sense of deep satisfaction or wonder at their movement having at last scaled the commanding heights of Britain's political structure. In any case the prime minister, Ramsay MacDonald, deflated them by delivering a sharp lecture on the need for punctuality. After this uninspiring start, some way down the agenda there came the unresolved question of the merger.

The cabinet agreed that the Chancellor of the Exchequer, Philip Snowden, should circulate a memorandum for discussion.[12] The memorandum concentrated on the question of selling the government's shares in Anglo–Persian, which an official statement by the Labour party had already condemned as contrary to public policy.[13] Inchcape, as government director, had already been instructed to find out from Shell what it would offer for the shares in the event of a sale: so much for Watson's assumption that they would automatically be offered to Burmah Oil. Shell had quoted a satisfactory figure; yet the Treasury felt convinced that Shell – which it described as "a none too scrupulous combine" – would not hesitate to recoup its outlay from higher prices charged to the British consumer.

There was also the employment aspect, in the deepening recession. The Shell directors had already made it clear that Anglo–Persian's unprofitable Llandarcy refinery would have to be closed down, and the next target for extinction would inevitably be the Scottish shale industry, then under Anglo–Persian's wing but operating at a loss. A further question-mark hung over Anglo–Persian itself. Its share values had fallen sharply as oil prices had declined since 1922 and as disturbing rumours had circulated about its failure to discover new fields outside Persia. But its assets were well covered, so that the government had no grounds for selling out. On the contrary, a clear statement of refusal to sell would end the uncertainty that had so damaged the company's interests.

At the next cabinet meeting, ministers accepted the Treasury memorandum and agreed to retain the government holdings, thereby defeating the merger proposal. Watson was very put out by this decision, passed on to him by Snowden on 29 January, especially when he heard from an unnamed friend that the cabinet had reached it after little or no serious discussion. He therefore resolved to get the decision reversed by every means possible. Meanwhile he could do no more than prepare a long statement for the press denying, for instance, that both Burmah Oil and Shell had many foreign shareholders, and seeing that other misleading allegations in responsible newspapers were refuted.

So matters rested while the minority Labour government remained shakily in power until November 1924. Then another general election returned the Conservatives with a substantial majority. During the immediate post-election doldrums, the London correspondent of a New York paper, said to have been inspired by Standard Oil, reported that the new government contemplated selling the Anglo–Persian holdings. There was at once a rash of articles in the British press, and Greenway wrote to ask Baldwin as prime minister if the rumours were true. Without taking the matter to the cabinet, Baldwin announced that they were not.[14]

Despite his earlier favourable reactions, neither he nor Cunliffe-Lister (who had changed his name from Lloyd-Greame as a consequence of an inheritance) at the Board of Trade, had any desire to raise such a contentious issue so early in what seemed certain to be a full term of office. Of the other ministers, only Winston Churchill really counted: but as Chancellor of the Exchequer he found himself immersed in too many other complex problems to welcome its reopening.

Neither Watson nor Cohen could at that time do any personal lobbying as they were both in the east. Yet Cohen was convinced that they would be able to find a favourable pretext for renewing their efforts next summer or autumn. "If we succeed," he told Watson with some exaggeration, "we shall have a greater influence on the future history of the British empire than any single individual has in recent generations." Yet 1925, and the years following the return to the gold standard, with consequent deflationary pressures in Britain, really belonged to a different and harsher world. In the event, as an official historian has argued, to the oil companies involved the outcome of this too long drawn-out episode made little practical difference either way.[15]

The closer marketing links of Shell with Burmah Oil and also with Anglo–Persian in the later 1920s fall outside the period here under review; yet they left Burmah Oil still facing the problem of the limited reserves in the oilfields of Burma and India. While Anglo–Persian steadily took off into a truly world-wide organization, Burmah Oil was virtually confined to oilfields that were in the main beyond their peak yield. So the company had reached a distinct turning-point in its history by the closing months of 1924: by coincidence, this was just fifty years after the initial meeting of David Cargill and Kirkman Finlay had prepared the way for creating what later became the first Burmah Oil Company.

Burmah Oil had developed sufficient technical and financial strength to sponsor developments in Persia and Iraq which were to become infinitely greater than its own relatively modest undertakings in Burma and Assam. There may seem to be some irony in the fact that its directors were capable at one and the same time of looking upon the Anglo–Persian Oil Company as being virtually a wholly-owned subsidiary, for which they accepted very real responsibility and in whose affairs they participated as of right for many years, while acquiescing in a comparatively minor role for their own pioneering company. The fact is, however, that the high priority attached to the production of fuel oil for the Royal Navy required that Anglo–Persian's resources should be developed at a rate that would have been completely unrealistic in terms of normal commercial operations. It was against this background that those who were seconded to the task of providing Anglo–Persian's top management soon found themselves in charge of an organization developing so rapidly that they could no longer respond to the helm of Burmah Oil.

It was in these circumstances that Burmah Oil itself came increasingly to be seen primarily as an investment company, with important holdings in Anglo–Persian and – after 1928 – in Shell. The period from 1924 on was to see a rare succession of vicissitudes. Falling output from declining reserves was to be followed in 1942 by the voluntary destruction of the Burmese oilfields so that they could be denied to the Japanese invaders. After the war, accords with newly independent host governments were to lead in the course of time to the cessation of the company's traditional operations in Burma and India.

Not until the 1960s did it seek to re-assert its operational independence, when it began to diversify geographically and to widen the range of its activities. This process was under way when, in the 1970s, the oil industry of the middle east – which Burmah Oil had done so much to help establish – was rocked to its very foundations by the wars and tensions between Israel and its Arab neighbours. In the process, the foundations of Burmah Oil all but crumbled. As in 1924, the company came at the end of 1974 to another denouement in which, once again, its future course was to be determined by political decisions taken at the highest levels. But, in Kipling's words, that is another story.

Notes

1 The proposals of 1921–2 and their consequences are documented in PRO POWE 33/92. The ambivalent attitude of the Treasury can be seen in PRO T 161/142/S 12612, 31 July 1921–January 1922.
2 PRO FO 371/7027: Watson's memo of 29 July 1921.
3 Ibid., Watson's two memos dated 31 October 1921.

4 PRO CAB 24/132: memo by minister in charge of petroleum (Lloyd-Greame), 6 January 1922. The hostile views of the Admiralty are in a memo of deputy secretary addressed to Admiralty Board, 8 February 1922, PRO ADM 167/66.

5 PRO CAB 24/137: report of Committee on Proposed Amalgamation (S. Baldwin), 23 June 1922.

6 PRO CAB 23/30, cabinet 40(22), item 5, 20 July 1922.

7 Churchill's brief "career" as an oil lobbyist and would-be oil company director is touched upon in M. Gilbert, *Winston S. Churchill: V 1922–1939* (1976), pp. 8–16.

8 His own position in the affair caused him concern throughout. See M. Gilbert, *Winston S. Churchill V, Companion Volume, Part I, Documents*: "The Exchequer Years 1922–9" (1979) Churchill to Mrs Churchill, 15 August 1923, pp. 54–5. His important memo of 20 November 1923 on the whole affair is in ibid., pp. 68–9.

9 PRO ADM 116/3452 for Churchill's letter and memo of 7 November 1923. Questions (undated) raised by Admiralty officials and their thumbs down to the whole scheme are in PRO ADM 1/8658/55.

10 S. Roskill, *Naval Policy Between the Wars: I The Period of Anglo–American Antagonism 1919–1929* (1968), p. 32.

11 PRO CAB 23/46, cabinet 4 (24), item 11, 14 January 1924, referring to Amery's memo CP 20/24 (in CAB 24/164) of 10 January 1924. See J. Barnes and D. Nicholson (eds) *The Leo Amery Diaries I, 1896–1929* (1980), p. 363 (cf. pp. 346–7).

12 PRO CAB 23/47, cabinet 7(24) and 8(24), 23 and 28 January 1924, both items 11, referring to Treasury memo CP 32/24 (in CAB 24/164) of 26 January 1924.

13 R. Page Arnot, *The Politics of Oil* (1924), p. 52.

14 The brief official statement from 10 Downing Street is in *The Times*, 20 November 1924.

15 Payton Smith, *Oil, A Study of War-Time Policy and Administration*, pp. 13–14.

Appendices

Principal Characters

Abbreviations: (BOC = Burmah Oil Company; APOC = Anglo–Persian Oil Company; FF & Co. = Finlay Fleming & Co., Rangoon; RS & Co. = R. G. Shaw & Co., London; SW & Co. = Shaw Wallace & Co., Calcutta. D = *Dictionary of National Biography*; W = *Who Was Who*, under year of death).

ADAMSON, Robert William (1856–1921). Assistant FF & Co. 1879; general manager 1902–4. Director BOC (Glasgow) 1904–18.

ASHTON, Hubert Shorrock (1862–1943). Assistant, SW & Co. c. 1886; partner 1888. Partner RS & Co. 1907. Director Assam Oil Company 1907. London director BOC 1907; director 1915–43. Retired from executive work 1919.

BARNES, Sir Hugh Shakespear (1853–1940) KCSI 1903. Lt.-Gov. Burma 1903–5; member of Council of India, London 1905–13. London director BOC 1913. Director APOC 1909–40 (W).

CADMAN, Sir John (1877–1941). KCMG 1918; Lord Cadman 1937. Professor of Mining, Birmingham University 1908–20. Petroleum adviser, Colonial Office. Member, Slade delegation to Persia 1913–14. Director, Petroleum Executive 1917–21. Joined APOC 1921; director 1923; deputy chairman 1925; chairman 1927–41. (D, W.)

CAMPBELL, Andrew (1868–1941). Chemist, Dunneedaw refinery 1889; works manager 1893–1914. Advisory chemist to BOC and APOC 1914. Director, National Oil Refineries Ltd., 1920–25.

CARGILL, David Sime (1826–1904). Assistant, William Milne & Co., Colombo 1844–61; partner, Glasgow 1861. Director, Rangoon

Oil Company c. 1872. Acquired Rangoon Oil Company (wound up 1876) and ran it as a solely owned enterprise 1876–86. Chairman BOC 1886–1904.

CARGILL, Sir John Traill (1867–1954). Bart. 1920. Assistant FF & Co. 1890–3; BOC Glasgow Office 1893. Liquidator in BOC's reconstruction 1902; director 1902; chairman 1904–43. Director APOC 1909–43. (W.)

COHEN, Sir Robert Waley (1877–1952) KBE 1920. Joined Shell Transport & Trading Company 1901; director 1906; managing director 1907. Director, Asiatic Petroleum Company 1906 and of Anglo–Saxon Petroleum Company 1907. Group managing director, Royal Dutch–Shell, 1921. (D, W.)

CRAIG, Edward Hubert Cunningham (1874–1946). With Geological Survey of India 1896–1903; government geologist, Trinidad 1903–7. Geologist BOC 1907–12; geological adviser BOC 1912–46; to APOC 1912–26. Geological adviser, Petroleum Executive 1917–18. (W.)

CURRIE, Captain George William (1874–1960). Tanker captain BOC 1899; marine superintendent 1909–33.

D'ARCY, William Knox (1849–1917). Solicitor. Pursued legal, pastoral and mining interests in Queensland, Australia 1866–89. Returned to England 1889. Secured oil concessions from Shah of Persia 1901. Chairman, First Exploitation Company 1903. Director, Concessions Syndicate Ltd. 1905. Director BOC 1909–17. Director APOC 1909–17. (W.)

DETERDING, Sir Henri Wilhelm August (1866–1939) KBE 1920. Joined Royal Dutch Oil Company, 1896. Managing director, Asiatic Petroleum Company 1903; general managing director, Royal Dutch and director, Shell Transport & Trading Company 1907–36. (W.)

DEWHURST, Thomas (1881–1973). Formerly lecturer in geology, Queen's University, Belfast. Geologist BOC 1910; senior geologist BOC (Rangoon) 1916–22; chief geologist BOC (London) 1922–38. Geological adviser APOC; geological adviser BOC 1938–61.

EVES, Sir Hubert Heath (1883–1961) Kt. 1946. Assistant, FF & Co. 1909 (Formerly with Bulloch Bros.) BOC general manager in India 1914–21. Joined APOC 1921; director 1924–53: deputy chairman 1941–50. (W.)

FINLAY, Sir Campbell Kirkman (1875–1937) Kt. 1912. Son of Kirkman Finlay (*q.v.*). Assistant, FF & Co. 1897–1904; general manager 1904–12. In charge, BOC London Office 1912; London director 1913; Director, BOC, 1914–37, Director, APOC, 1912–18. War service with Ministry of Munitions 1915; Ministry of Food 1916–20. (W.)

FINLAY, Kirkman (1847–1903). Assistant, Galbraith Dalziel & Co., Rangoon (agents of Rangoon Oil Company). Met David Cargill (*q.v.*) 1874. Co-founder FF & Co. 1876. Returned to Glasgow 1879. Company secretary BOC 1889–91; director 1891; managing director 1900. Set up BOC's London Office 1891.

FLEMING, Matthew Tarbett (1851–1913). Formerly Assistant, British India Steam Navigation Company, Rangoon office. Co-founder FF & Co. 1876. Returned to Glasgow c. 1877. Director BOC 1886–1913.

GALBRAITH, James (1818–85). Clerk P. Henderson & Co., shipowners 1844; partner 1848; senior partner 1867. Managing director, Irrawaddy Flotilla (and Burma Steam Navigation) Company 1865–85. Chairman and managing director, Rangoon Oil Company 1871–76.

GALBRAITH, James (1834–1919). Senior partner Galbraith Reid & Co., Glasgow and of Galbraith Dalziel & Co., UK and Rangoon agents respectively of Rangoon Oil Company 1871–76. Partner James Struthers & Co., coal merchants and oil distillers, Stand, near Airdrie, to 1876.

GARROW, Duncan (1878–1931). Assistant FF & Co., 1903; general manager 1912–14. Joined APOC 1914; director 1914–24; managing director 1919.

GILLESPIE, John (1865–1944). Partner, A. Gillespie & Son, engineers, Glasgow. Consulting engineer BOC 1897–1928. Helped to plan pipelines in Burma and Persia. Collaborated with Andrew Campbell (*q.v.*) in planning Abadan refinery.

GREENWAY, Sir Charles (1857–1934). Bart. 1919; Lord Greenway 1927. Assistant SW & Co. 1893; senior partner 1907. Partner RS & Co. 1907. London director BOC 1907. Director APOC 1909–34; managing director 1910–19; chairman 1914–27. Retired 1927. (W.)

HAMILTON, James (1860–1920). Clerk, BOC Glasgow 1886; company secretary 1891–1905. "Manager" in Glasgow office 1906–19; director BOC 1905–20. Director APOC 1909–20.

HOLLAND, Sir Thomas Henry (1868–1947). KCIE 1908, KCSI 1918. Assistant superintendent, Geological Survey of India 1890; director 1903–9. Professor of Geology and Mineralogy, Manchester University 1909–18. President, Burma Oilfields Committee 1908. Member, Royal Commission on Fuel and Engines 1912–13. Member for Commerce and Industry, Viceroy's Executive Council, India 1920–1. Rector, Imperial College of Science and Technology 1922–9. Principal and Vice-Chancellor, Edinburgh University 1929–44. (D, W.)

INNES, John (d. 1928). Partner, P. Henderson & Co. 1887. Managing director Irrawaddy Flotilla Company 1894–1927; chairman 1924. Director BOC 1897–1909.

JACOBS, Carl B. (d. 1916). American. General fields manager BOC 1902–15.

MINDON MIN King of Upper Burma (Ava), (reigned 1853–78). Decreed earth-oil a royal monopoly 1854.

MURRAY, Robert Alexander (d. 1937). Senior partner Brown Fleming & Murray, accountants, Glasgow 1889. Auditor BOC 1889–1920. Director BOC 1920–37.

PENNYCUICK, Alexander (1844–1906) CIE 1898. Formerly with Gladstone Wylie & Co., Rangoon. Assistant FF & Co. 1875; general manager 1879–1902. (W.)

REDWOOD, Sir Boverton (1846–1919). Kt. 1905; Bart. 1911. Leading oil consultant in UK to public bodies (including Admiralty, India Office and Port of London Authority). Consultant to BOC 1893; to W. K. D'Arcy (*q.v.*) 1901; to APOC 1909. Member, Oil Fuel Committee 1902–6; Royal Commission on Fuel and Engines 1912–13. Director of Technical Investigations, Petroleum Executive 1917–19. (W.)

REYNOLDS, George Bernard (1852–1928). Formerly with Indian Public Works Department. In charge of prospecting and drilling operations, northern (and later southern) Persia 1901–11. Made oil strike at Masjid-i-Sulaiman 1908.

RITCHIE, Sir Adam Beattie (1881–1957). Kt. 1923. Assistant FF & Co. 1906. BOC general manager in India 1911–12; general manager FF & Co. 1914–26. General manager Turkish (later Iraq) Petroleum Company 1926–29. (W.)

SAMUEL, Sir Marcus (1853–1927). Kt. 1898; Bart. 1903; Lord Bearstead 1921; Viscount 1925. Founder and first chairman. Shell Transport & Trading Company 1897. Lord Mayor of London 1902–3. Chairman Asiatic Petroleum Company 1903. Retired 1920. (D, W.)

SEIPLE, John S. From Lancaster, Ohio, USA. Joined BOC as driller 1897; general fields manager 1915–23. Relative of W. Seiple (*q.v.*).

SEIPLE, William. From Lancaster, Ohio, USA. Joined BOC 1890; general fields manager to 1902. Relative of J. S. Seiple (*q.v.*).

STRATHCONA AND MOUNT ROYAL, Lord (Donald Alexander Smith, 1820–1914). Kt. 1896; baron 1897. Chief commissioner, director and governor of Hudson Bay Company. Director Canadian Pacific Railway Company. High Commissioner for Canada in London 1897. Shareholder, Concessions Syndicate Ltd, 1905; chairman APOC 1909–14. (D, W.)

THIBAW King of Upper Burma, (reigned 1878–85). Received Kirkman Finlay (*q.v.*) 1879. Defeated in third Anglo–Burmese war 1885. Exiled to Ratnagiri, near Bombay; died there 1916.

WALLACE, Charles William (1855–1916). Co-founder SW & Co. 1886. Returned to become partner RS & Co. 1892. Director BOC 1902–15; offered but refused managing directorship. Director APOC 1909–16, managing director 1909–10; vice-chairman 1909–15.

WATSON, Robert Irving (1878–1948). In London Office BOC 1901–2 and 1912 onwards. Assistant FF & Co. 1902–12. London director BOC 1914; director BOC 1918–47; managing director 1920–47, chairman 1943. Director APOC 1918–47. Director Shell Transport & Trading Company 1929–47.

WHIGHAM, Gilbert Campbell (1877–1950). Assistant FF & Co. 1904. BOC general manager in India 1912–14. Joined BOC London Office 1919; London director 1919; director 1920–46; assistant managing director 1923–45. Director APOC 1925–46.

Burmah Oil Company – Financial Data
(£000s)

Year	Issued ordinary shares	Issued preference shares	Debentures	Total assets	Gross profit (earned in Rangoon)	Investment income	Total income	Net profit
1889	54	—	—	..	17	—	17	8
1890	54	—	—	..	18	—	18	9
1891	79	—	—	..	21	—	21	14
1892	87	—	—	..	24	—	24	17
1893	102	—	—	..	25	—	25	18
1894	120	—	—	..	24	—	24	16
1895	120	—	—	..	35	—	35	28
1896	168	—	—	..	82	—	82	75
1897	168	—	—	..	52	—	52	43
1898	200	100	50	..	62	—	62	45
1899	200	100	50	..	97	—	97	77
1900	200	100	50	..	163	—	163	135
1901	200	107	50	..	131	—	131	96

Year								
1902(a)	980	165	100	1,481	214	—	214	185
1903	991	165	500	2,125	332	—	332	265
1904	1,100	250	480	2,587	382	—	382	279
1905	1,100	300	460	3,096	316	—	316	202
1906	1,100	300	440	3,363	548	19	567	394
1907	1,100	325	378	3,886	726	33	759	607
1908	1,100	325	317	4,138	718	41	759	604
1909	1,270	1,000	280	4,568	607	33	640	552
1910	1,905	1,000	260	4,715	678	34	712	554
1911	1,905	1,000	240	4,655	497	13	510	357
1912	1,905	1,000	220	4,997	755	24	779	623
1913	1,905	1,000	—	5,211	962	42	1,004	843
1914	1,905	1,000	—	5,505	971	58	1,029	867
1915	1,905	1,000	—	5,692	935	57	992	743
1916	1,905	1,000	—	6,492	1,354	75	1,429	1,197
1917	1,905	1,000	—	8,191	2,306	161	2,467	2,225
1918	2,858	1,000	—	8,901	2,849	215	3,064	2,734
1919	2,858	1,000	—	10,940	4,344	285	4,629	4,108
1920	5,144	1,000	—	12,332	3,966	320	4,286	3,738
1921	5,151	4,000	—	15,123	2,354	282	2,636	1,903
1922	5,151	4,000	—	15,217	2,428	358	2,786	2,116
1923	5,151	4,000	—	14,814	2,453	439	2,892	2,204
1924	5,151	4,000	—	15,482	2,493	387	2,880	2,257

Source: Company Records
Notes: .. Not available
— Nil
(a) Company reconstructed 1902

Crude Oil Production and Refinery Throughput Data

Year	Total crude oil production – Burma (a) (000 barrels)	Burmah Oil Company refinery throughput (000 barrels)	Gross profit per barrel of refinery throughput (pence)	Net profit per barrel of refinery throughput (pence)
1795	25	—	—	—
1826	37	—	—	—
1855	46	—	—	—
1873	53	—	—	—
1884	44	—	—	—
1885	45	—	—	—
1886	50	48	—	—
1887	57	48	—	—
1888	65	50	—	—
1889	72	60	28.3	13.3
1890	110	64	28.1	14.7
1891	144	80	26.7	17.9
1892	211	103	23.1	16.5
1893	251	118	21.2	15.5
1894	269	139	17.2	11.7
1895	328	163	21.6	17.0
1896	377	194	42.1	38.5
1897	439	249	20.7	17.2
1898	513	310	20.0	14.4
1899	737	396	24.4	19.5
1900	936	591	27.5	22.7
1901	1,215	758	17.2	12.6
1902	1,385	969	22.1	19.1
1903	2,133	1,361	24.4	19.4
1904	2,898	2,038	18.7	13.7
1905	3,550	2,771	11.4	7.3
1906	3,441	2,933	18.6	13.4
1907	3,722	3,217	22.6	18.9
1908	4,334	3,430	20.9	17.6
1909	5,760	3,901	15.6	14.2
1910	5,288	3,960	17.1	14.0

Production Data

Year	Total crude oil production – Burma (a) (000 barrels)	Burmah Oil Company refinery throughput (000 barrels)	Gross profit	Net profit
			per barrel of refinery throughput	
			(pence)	(pence)
1911	5,556	3,970	12.5	9.0
1912	6,133	4,001	18.9	15.6
1913	6,822	4,506	21.3	18.7
1914	6,366	4,554	21.3	19.0
1915	7,057	4,509	20.7	16.5
1916	7,294	4,652	29.1	25.7
1917	6,820	4,676	49.3	47.6
1918	6,871	4,687	60.8	58.3
1919	6,898	4,650	93.4	88.3
1920	6,625	4,668	85.0	80.1
1921	7,044	4,588	51.3	41.5
1922	6,758	4,760	51.0	44.4
1923	6,785	5,037	48.7	43.8
1924	6,692	4,951	50.4	45.6

Sources: Pre-1886 estimates based on F. Noetling, *The Occurrence of Petroleum in Burma* (1898). 1886 onwards, company records.

Notes: (a) Pre-1886 figures exclude oil raised in Arakan, etc.
Post-1886. All companies and well-owners in Burma.

Index